Optimization Algorithms – Classics and Recent Advances

Edited by Mykhaylo Andriychuk and Ali Sadollah

Published in London, United Kingdom

Optimization Algorithms - Classics and Recent Advances
http://dx.doi.org/10.5772/intechopen.111098
Edited by Mykhaylo Andriychuk and Ali Sadollah

Contributors
Abderrahman El Boukili, Alireza Rezvanian, Ali Sadollah, Hicham Lekbich, Hossein Hatami, Isaac Wilfried Sanou, Kapil Choudhary, Marvin Chandra Wijaya, Mehra Esfandiari, Moulay Brahim Sedra, Mykhaylo I. Andriychuk, Najim Mansour, Nasrin Jokar, Nissara Sirasuntorn, Oladimeji Samuel Sowole, Ramadan Abdelaziz, Roland Redon, Ruangvarin Intarawong Sararnrakskul, S. Mehdi Vahidipour, Shima Aghamirzadeh, Siddique Kamran, Stéphane Mounier, Tahir Toghrai, Thorranin Pawaputanon Na Mahasarakham, Xavier Luciani, Xiaofeng Yang, Yoshifumi Ogami

First published in London, United Kingdom, 2024 by IntechOpen
IntechOpen is the global imprint of INTECHOPEN LIMITED, registered in England and Wales, registration number: 11086078, 167-169 Great Portland Street, London, W1W 5PF, United Kingdom

British Library Cataloguing-in-Publication Data
A catalogue record for this book is available from the British Library

Additional hard and PDF copies can be obtained from orders@intechopen.com

Optimization Algorithms - Classics and Recent Advances
Edited by Mykhaylo Andriychuk and Ali Sadollah
p. cm.
Print ISBN 978-1-83769-180-7
Online ISBN 978-1-83769-181-4
eBook (PDF) ISBN 978-1-83769-182-1

For EU product safety concerns:
IN TECH d.o.o., Prolaz Marije Krucifikse Kozulić 3, 51000 Rijeka, Croatia,
info@intechopen.com or visit our website at intechopen.com.

Meet the editors

Prof. Andriychuk received an MSc in Computational Mathematics from Lviv National University, Ukraine, a Ph.D. in Application of Computational Techniques from Kyiv National University, Ukraine, and a DSc in Mathematical Modelling from Lviv Polytechnic National University (LPNU), Ukraine, in 1976, 1987, and 2015, respectively. He has been with the Pidstryhach Institute for Applied Problems of Mechanics and Mathematics (IAPMM), Ukraine, for 48 years. Currently, he is the Head of the Department at IAPMM and a professor at LPNU. He has published about 200 papers on diffraction and antenna synthesis theory, optimization methods, and nonlinear equations in scientific journals and international conference proceedings. He is also the author of two books on antenna theory. He has been an IEEE Member since 1995 and IEEE Senior Member since 2003.

Dr. Ali Sadollah received his Ph.D. in Mechanical Engineering, from the University of Malaya, Kuala Lumpur, Malaysia, in 2013. He has served as a postdoctoral research fellow at different universities such as Korea University and Nanyang Technological University, Singapore. He was also a guest assistant professor at Sharif University of Technology, Tehran, Iran. Currently, he is an assistant professor at the University of Science and Culture, Tehran, Iran, as well as the vice president of research and technology and head of the international cooperation office. He is ranked among the top 2 and 1 percent of the world's highly cited scholars, according to Stanford University, California, USA. He is the inventor of two metaheuristic optimization methods: the water cycle algorithm and the neural network algorithm. His research interests include algorithm development, optimization and metaheuristics, applications of soft computing methods in engineering, computational solid mechanics, finite element method, and cold roll forming process. Dr. Sadollah has published four books and contributed to more than 150 scientific studies. He has been invited to participate in numerous international and national events.

Contents

Preface

Optimization algorithms are a set of algorithms that allow finding the best possible solution to a stated problem. Typically, these algorithms aim to identify the solution that maximizes or minimizes a given objective function. Such algorithms are used comprehensively for solving many problems related to pure science and engineering applications.

This book is devoted to the development of the optimization algorithm theory and the application of the different types of optimization algorithms to solving problems in theoretical physics, signal processing, electromagnetism, the food industry, ecology, and more. The book consists of two sections. Section 1 contains six chapters that discuss the development of the algorithm theory and present examples of using optimization algorithms for solving practical engineering problems.

In Chapter 1, T. P. N. Mahasarakham et al. deal with abstract objects in specific spaces that can be applied to the construction of a generalized algorithm theory. They introduce the tensor products of ternary semimodules (abstract data) over ternary semifields (specific spaces) and the n-fold tensor products for a general case. The chapter establishes the universal mapping property of n-fold tensor products and outlines conditions for maintaining the flatness of ternary semimodules through tensor products. These introduced concepts enable the exploration of various properties of ternary semimodules over ternary semifields and serve as foundational tools for the development of generalized algorithms tailored to specific datasets.

A series of algorithms in physics is based on partial differential equations, particularly the Korteweg-deVries (KdV) equation. In Chapter 2, A. El Boukili et al. present a basic survey of the two-dimensional KdV model. They examine the various symmetries of systems and explore the concept of integrability through two different methods: Hamiltonian formalism and the identification of conserved quantities. By introducing the concept of q-deformation, the authors establish the corresponding q-deformation integrable model and investigate the resulting system's integrability. The chapter presents the algebraic structure of the system and studies integrability using the notion of a Lax pair.

In Chapter 3, I. W. Sanou et al. apply canonical polyadic decomposition (CPD) or parallel factor analysis in signal processing for the decomposition of multi-way arrays. This approach is applied when it is necessary to add constraints to provide a quick converge of the optimal solution. In contrast to classic CPD, the authors focus on online CPD. In this context, the number of relevant factors is usually unknown and can vary with time. The authors propose two algorithms to compute online CPD based on sparse dictionary learning. The chapter concludes with an application example in environmental sciences and an evaluation of the performances of the proposed approaches on real data.

In Chapter 4, M. Andriychuk focuses on optimization problems with restrictions of the objective function. Such problems arise in antenna synthesis applications, necessitating the resolution of nonlinear integral equations that correspond to the respective functionals of the synthesis problem. The nature of the reduced equations depends on the operator used for calculating the antenna's directivity pattern. The chapter considers two types of restrictions: where the amplitude of the objective function is prescribed (the problem of phase synthesis), and where restrictions are imposed on the phase of the objective function (the problem of amplitude synthesis). The method of successive approximations is effective for solving these reduced equations. Notably, these equations exhibit non-uniqueness in their solutions. The chapter corroborates the accuracy of the analytical approach using numerical data. The advantage of the proposed approach is a fast convergence of iterative procedures employed to solve the obtained equations.

In Chapter 5, K. Choudhary applies Ant Colony Optimization (ACO) for agricultural price forecasting. The chapter explores the innovative use of ACO-based models, backed by insightful case studies that demonstrate their practical applicability. The chapter examines the synergy between ACO and agriculture price forecasting with a focus on the unique challenges and opportunities presented by this domain. It also discusses future prospects of ACO in this field. Furthermore, the chapter defines emerging trends, potential areas for improvement, and avenues for further research. Overall, the chapter advocates for collaborative efforts among researchers, practitioners, and policymakers to fully leverage the capabilities of ACO-based models.

In Chapter 6, A. Ramadan applies ACO to water resource management. The proposed approach allows for maximizing the use of available resources while minimizing negative impacts on the environment. To attain the stated goal, two approaches are used: global and local optimization. The data obtained suggests that ACO is a flexible and powerful tool that can handle various types of hydrological models. The chapter employs different performance measures and case studies to comprehensively compare these techniques in water resource management. The obtained results are useful for practitioners in choosing the appropriate optimization technique for water resource management.

Section 2 is devoted to the application of optimization algorithms in information technology, including computer implementation. Chapters in this section discuss problems of computer parameter optimization, multimedia data, and database processing, optimization of recommender systems, transportation improvement, and application of ant colony algorithms for dynamic optimization.

In Chapter 7, O. Yoshifumi and S. Kamran apply Cartesian cell registration to the problem of reducing both computation time for registration and computer memory. The authors implement this procedure within an existing FORTRAN code previously utilized for dilute fluid flows and multiphase flow calculations. They review the Cartesian grid system method, highlighting its disadvantages, which necessitate the application of more effective methods such as the Cartesian cell registration method. The authors explain the peculiarity of the Cartesian cell registration method via its implementation within a FORTRAN-based algorithm. In addition, they present improvements to the conventional algorithm aimed at accelerating processing speed.

The chapter showcases the application of the direct simulation Monte Carlo (DSMC) method utilizing the proposed algorithm to address the challenge of two-dimensional Argon flows around a circular cylinder.

In Chapter 8, M. C. Wijaya investigates the reduction of kernel mechanism processing time in the multimedia authoring tool. The approach is applicable to both visual media (images, video, animation, text, etc.) and non-visual media (sound, music, etc.). In this approach, the multimedia documents are presented in a multimedia programming language. The core process in multimedia authoring tools (the kernel mechanism) consists of two essential stages: modeling and verification. Verification involves both time-computation and spatial-temporal verification. The time required for verification in a multimedia authoring tool depends on the object or media being processed. The chapter highlights the exponential relationship between processing time and media volume, which poses a significant challenge for multimedia authoring tools. The proposed incremental algorithm can be applied to the verification process to reduce processing time. Experimental results demonstrate that this approach can yield up to a 75% reduction in processing time, showcasing its effectiveness in enhancing the efficiency of multimedia document verification.

Chapter 9 by X. Yang presents a comprehensive review of algorithms used in database access and optimization strategies. It compares two principal access methods based on SQL and NoSQL database techniques, highlighting their respective advantages and disadvantages. The comparison is explained through the lens of Oracle Corporation's database techniques, known for their depth-oriented principle, and Amazon's Dynamo database, characterized by its document-oriented structure. The chapter also introduces a general optimization technique that leverages indexing, cashing, and metrics to enhance database access efficiency. It analyzes and compares various tools impacting the overall optimization process. Furthermore, the chapter proposes a concept for arranging access optimization for specific applications. The main focus is on formulating database performance metrics effectively, utilizing the resource description framework model and its internal tools, and implementing specific measures and evaluating their efficacy.

In Chapter 10, N. Jokar et al. study recommender systems and aim to personalize user recommendations by using the textual information from published papers. Usually, studies of this type are based on textual elements from the literature, including title, abstract, keywords, publisher, author, and other similar items. Since the quantity of papers is increasing day by day, recommender systems must be able to process huge volumes of data according to the user's needs. To achieve this goal, the authors apply big data tools, which facilitate parallel processing to deliver relevant recommendations.

In Chapter 11, O. S. Sowole reviews the application of minimizing algorithms for solving vehicle routing problems (VRPs). Such problems appear in logistics, transportation, and supply chain management. The VRP involves searching for the optimal set of routes for a fleet of vehicles that serves a given set of customers while minimizing the total travel distance and maximizing total customer needs. The examples in the chapter illustrate how different search algorithms have been applied to solve VRPs in transportation applications like vehicle routing for package delivery, waste collection, and emergency response. The chapter analyzes input data, which defines the performance of search algorithms while solving the VRP relative to problem size, complexity, and algorithm parameters. The capacitated VRP is formulated, and the method of solving is explained using a case example of four customers. The chapter explains how a series of genetic algorithms can be applied successfully to solve the VRP. It also provides recommendations for choosing the optimal algorithms for specific cases and highlights future use cases for the discussed algorithms.

In Chapter 12, A. Rezvanian et al. propose an overview of ant colony optimization algorithms, which are applied to dynamic optimization problems. The analysis of the different kinds of optimization problems demonstrates that ant colony optimization can be an effective tool to solve most of them. As shown in the recent computational practice, solving the synthetic dynamic optimization problems by the ant colony algorithms allows getting a variety of real-world solutions closely. Since the exact algorithms are impractical in dynamic environments, stochastic optimization techniques have become popular. Related to these techniques, the ant colony optimization becomes favorable for solving dynamic optimization problems owing to its capacity to adapt to quickly changing situations. The presented biometric analysis confirms the importance of ant colony optimization in solving different life, social, communication, technical, and pure science problems.

It is our great pleasure to thank all authors for their excellent chapters. In addition, I express my sincere appreciation to Ms. Karmen Đaleta for her professional support and patience during the publishing process.

Mykhaylo Andriychuk
Pidstryhach Institute for Applied Problems of Mechanics and Mathematics,
National Academy of Sciences of Ukraine,
Ukraine

Lviv Polytechnic National University,
Lviv, Ukraine

Ali Sadollah
Department of Mechanical Engineering,
University of Science and Culture,
Tehran, Iran

Section 1

Algorithm Theory and Applications

Chapter 1

Tensor Products of Ternary Semimodules over Ternary Semifields

Thorranin Pawaputanon Na Mahasarakham,
Ruangvarin Intarawong Sararnrakskul and Nissara Sirasuntorn

Abstract

We introduce tensor products of ternary semimodules over ternary semifields and the n-fold tensor products for general case. We prove the universal mapping property of the n-fold tensor products. Moreover, we later provide a condition for preserving the flatness of ternary semimodules with tensor products of ternary semimodule homomorphisms.

Keywords: tensor product, ternary semimodule, ternary semifield, ternary algebra, abelian groups

1. Introduction

The subject of ternary algebra was started by D. H. Lehmer [1]. He studied some concepts of ternary systems called triplexes, which obtained the generalization of abelian groups. Afterward, Los [2] mentioned this algebraic structure studied by Banach and showed an example of a ternary semigroup, which is not a semigroup. In 1971, W. G. Lister [3] studied the abstract structure of a ternary ring, which is a ternary product of abelian groups. In 2003, R. Intarawong [4] studied and provided the universal mapping property of tensor products of modules over semifields. T. K. Dutta and S. Kar [5] studied ternary semiring and gave some properties of ternary semifields. In addition, H. J. M. Al-Thani [6] studied the flat semimodules, which is constructed by the exact sequences of semimodules and tensor products of semimodule homomorphisms.

The universal mapping property in the branch of modules over rings gives that there exists a unique group homomorphism from $M \otimes_R N$ to a module A, which is composed with a bilinear map $M \times N \to M \otimes_R N$ is a bilinear map $M \times N \to A$. The structure of the ternary semimodules also leads us to derive a similar result.

The goal of this research is to investigate some properties of ternary semimodules over ternary semifields. We also study the universal mapping property of tensor products of ternary semimodules and investigate some types of ternary semimodules defined by tensor products of homomorphisms on ternary semimodules and exact sequences.

 IntechOpen

2. Preliminaries

The following familiar definitions and theorems from Refs. [7–9] regarding the notion of free abelian groups are needed to define tensor products of ternary semimodules over ternary semifields.

A ternary semifield K is a system $(K, +, \cdot)$ with a binary operation $(+)$ and a ternary operation $(\cdot)$ satisfying the following conditions for all $u, v, w, s, t \in K$.

1. $(K, +)$ is a commutative semigroup with identity 0,

2. $(u \cdot v \cdot w) \cdot s \cdot t = u \cdot (v \cdot w \cdot s) \cdot t = u \cdot v \cdot (w \cdot s \cdot t)$,

3. $(u + v) \cdot w \cdot s = u \cdot w \cdot s + v \cdot w \cdot s$,

4. $u \cdot (v + w) \cdot s = u \cdot v \cdot s + u \cdot w \cdot s$, and.

5. $u \cdot v \cdot (w + s) = u \cdot v \cdot w + a \cdot v \cdot s$.

6. $u \cdot v \cdot w = v \cdot u \cdot w = w \cdot v \cdot u = u \cdot w \cdot v$,

7. $\exists 0 \in K \ \ \forall u, v \in K, \ \ 0 + u = u$ and $0 \cdot u \cdot v = u \cdot 0 \cdot v = u \cdot v \cdot 0 = 0$, and.

8. $\forall u \in K \setminus \{0\} \ \ \exists v \in K \ \ \forall t \in K, \ \ u \cdot v \cdot t = v \cdot u \cdot t = t \cdot u \cdot v = t \cdot v \cdot u = t$.

(An element v is called an *inverse* of u. In addition, the inverse of u is unique and u^{-1} denotes the inverse of u.)

For convenience, let uvw denote $u \cdot v \cdot w$ for all $u, v, w \in K$.

Let $(R, +, \cdot)$ be a commutative ring. For a nonempty set S of R, a subring S of R is said to be a *positive cone* if $S \cup (-S) = R$ and $S \cap (-S) = \{0\}$. $-S$ is called a *negative cone* of S.

It is clear that every negative cone of a cone S of R is a ternary semiring. For example, $\mathbb{Z}_0^-$ is a natural example of a ternary semiring. Moreover, $\mathbb{Q}_0^-$ and $\mathbb{R}_0^-$ become ternary semifields.

A group $\mathcal{F}$ is a *free abelian group* if $\mathcal{F}$ is an abelian group and for every nonzero element g of $\mathcal{F}$, there exist unique nonzero integers $\alpha_1, \alpha_2, \ldots, \alpha_n$ and unique distinct $x_1, x_2, \ldots, x_n$ in $X \subseteq \mathcal{F}$ such that $g = \alpha_1 x_1 + \alpha_2 x_2 + \cdots + \alpha_n x_n$. We sometimes call X a *basis* for $\mathcal{F}$.

For a nonempty set X, let

$$\mathcal{FA}(X) = \{f : X \to \mathbb{Z} \mid \exists F \subseteq X \text{ such that } |F| < \infty \text{ and } f(x) = 0 \text{ for all } x \in X \setminus F\}. \quad (1)$$

Define $+$ on $\mathcal{FA}(X)$ by for any $f, g \in \mathcal{FA}(X)$,

$$(f + g)(x) = f(x) + g(x) \quad \text{for all } x \in X. \quad (2)$$

Then, $(\mathcal{FA}(X), +)$ is an abelian group. For any $x \in X$, define $f_x : X \to \mathbb{Z}$ by

$$f_x(y) = \begin{cases} 1, & \text{if } x = y, \\ 0, & \text{otherwise.} \end{cases} \quad (3)$$

Then, $f_x \in \mathcal{FA}(X)$ for all $x \in X$. A group $(\mathcal{FA}(X), +)$ is a free abelian group on $\{f_x | x \in X\}$ $(\subseteq \mathcal{FA}(X))$. We sometimes say instead that $\mathcal{FA}(X)$ is a free abelian group on X.

Let K be a ternary semifield. A *left K-ternary semimodule* or *left ternary semimodule over K* is an additive abelian group M together with a function from $K \times K \times M$ into M, defined by $(k_1, k_2, m) \mapsto k_1k_2m$ called ternary scalar multiplication, which satisfies the following conditions for all $m, m_1, m_2 \in M$ and $k_1, k_2, k_3, k_4 \in K$,

1. $k_1k_2(m_1 + m_2) = k_1k_2m_1 + k_1k_2m_2$,

2. $k_1(k_2 + k_3)m = k_1k_2m + k_1k_3m$,

3. $(k_1 + k_2)k_3m = k_1k_3m + k_2k_3m$,

4. $(k_1k_2k_3)k_4m = k_1(k_2k_3k_4)m = k_1k_2(k_3k_4m)$.

A *right K-ternary semimodule* or *right-ternary semimodule over K* is defined similarly via a function $M \times K \times K$ into M and satisfies the obvious analogs of (1)–(4).

For convenience, we simply write ${}_KM$ $[M_K]$ as M is a left [right] ternary semimodule over a ternary semifield K.

From now on, unless specified otherwise, "K-ternary semimodule ${}_KM$ $[M_K]$" means "left [right] K-ternary semimodule M ." Moreover, "K-ternary semimodule" means "left K-ternary semimodule."

Example 1.1 Let $F[0,1] = \{f | f : [0,1] \to \mathbb{Q}_0^-\}$ with the operations $+$ and $\cdot$, defined by.

$$(f + g)(x) = f(x) + g(x) \text{ and } (\alpha \cdot \beta \cdot f)(x) = \alpha \cdot \beta \cdot f(x) \text{ for all } x \in [0,1], \qquad (4)$$

where $f, g \in F[0,1]$ and $\alpha, \beta \in \mathbb{Q}_0^-$. Then, $F[0,1]$ is a left $\mathbb{Q}_0^-$-ternary semimodule.

Example 1.2 If $n \in \mathbb{N}$ and $M_1,\ M_2,\ \ldots,\ M_n$ are ternary semimodules over a ternary semifield K; then, $M_1 \times M_2 \times \cdots \times M_n$ is a ternary semimodules over K under usual componentwise addition and scalar multiplication.

Let M be a left [right] ternary semimodule over a ternary semifield K. A *left [right] ternary subsemimodule* of M is a subset of M, which is itself a left [right] ternary semimodule over K with the addition and ternary scalar multiplication of M.

Let K and S be ternary semifields. An additive abelian group M is a (K, S)-*ternary bisemimodule* if M is a left K-ternary semimodule and also a right S-ternary semimodule, and $k_1k_2(ms_1s_2) = (k_1k_2m)s_1s_2$ for all $k_1, k_2 \in K, m \in M,$ and $s_1, s_2 \in S$. We write ${}_KM_S$ for a (K, S)-ternary bisemimodule M.

3. Universal mapping properties

For given ternary semimodules M and N, over a ternary semifield K, it is known from Example 1.2 that $M \times N$ is a ternary semimodule over K. To find another ternary semimodule over K, arising from M and N, which is different from $M \times N$, the tensor product of M and N is the case. The notion of free abelian groups plays a major role in constructing the tensor product of ternary semimodules over K.

Let M_K and $_KN$ be ternary semimodules over a ternary semifield K. For each $m \in M$ and $n \in N$, a function $f_{(m,n)} : M \times N \to \mathbb{Z}$ is defined by

$$f_{(m,n)}(x,y) = \begin{cases} 1, & \text{if}(x,y) = (m,n), \\ 0, & \text{otherwise} \end{cases} \tag{5}$$

for all $x \in M_K$ and $y \in {}_KN$. Then, $f_{(m,n)} \in \mathcal{FA}(M \times N)$ and $\mathcal{FA}(M \times N)$ is a free abelian group on a basis $\left\{f_{(m,n)} \mid m \in M,\ n \in N\right\}$.

Definition 1.3 Let M_K and $_KN$ be ternary semimodules over a ternary semifield K, and let $\mathcal{F}$ be the free abelian group on $M \times N$, that is, $\mathcal{F} = \mathcal{FA}(M \times N)$. Let L be the subgroup of $\mathcal{F}$ generated by elements of the following forms:

1.$f_{(m+\tilde{m},n)} - f_{(m,n)} - f_{(\tilde{m},n)}$,

2.$f_{(m,n+\tilde{n})} - f_{(m,n)} - f_{(m,\tilde{n})}$,

3.$f_{(m\alpha\beta,n)} - f_{(m,\alpha\beta n)}$

where $\alpha, \beta \in K$, $m, \tilde{m} \in M$, and $n, \tilde{n} \in N$. We call $\mathcal{F}/L$ the *tensor product* of M and N, and denoted by $M \otimes_K N$.

Note 1.4 If M_K and $_KN$ are ternary semimodules over a ternary semifield K. Then,

1. $(M \otimes_K N, +)$ is an abelian group.

2.$f_{(m,n)} \in \mathcal{F}$ for all $m \in M_K$ and $n \in {}_KN$.

Recall that $\mathcal{F} = \mathcal{FA}(M \times N)$ and $\mathcal{F}/L = \{f + L | f \in \mathcal{F}\}$. For any $m \in M$ and $n \in N$, we write $m \otimes n$ for $f_{(m,n)} + L$. Moreover, for all $f + L \in \mathcal{F}/L$,

$$f + L = \sum_{(m,n) \in F \subseteq M \times N} \alpha_{(m,n)} \beta_{(m,n)} (m \otimes n), \tag{6}$$

where $\alpha_{(m,n)}, \beta_{(m,n)} \in \mathbb{Z}$ and $|F| < \infty$, and we simply write $f + L$ as

$$\sum_{i=1}^{p} \alpha_{(m,n)} \beta_{(m,n)} (m_i \otimes n_i) \text{ or } \sum_{i=1}^{p} \alpha_i \beta_i (m_i \otimes n_i) \tag{7}$$

where $\alpha_i = \alpha_{(m_i,n_i)}, \beta_i = \beta_{(m_i,n_i)} \in \mathbb{Z}$, $m_i \in M$, and $n_i \in N$.

The following property can be derived directly from Definition 1.3.

Let M_K and $_KN$ be ternary semimodules over a ternary semifield K. Then,

1. $(m + m') \otimes n = (m \otimes n) + (m' \otimes n)$,

2.$m \otimes (n + n') = (m \otimes n) + (m \otimes n')$,

3.$m\alpha\beta \otimes n = m \otimes \alpha\beta n$,

4. $m \otimes 0 = 0 \otimes n = 0 \otimes 0 = 0$,

for all $\alpha, \beta \in K$, $m, m' \in M$ and $n, n' \in N$.

Let M_K and ${}_K N$ be ternary semimodules over a ternary semifield K, and A is an additive abelian group. A *middle linear map* (over K) from $M \times N$ to A is a function $\tau : M \times N \rightarrow A$ such that for all $m, \tilde{m} \in M$, $n, \tilde{n} \in N$, and $\alpha, \beta \in K$,

1. $\tau(m + \tilde{m}, n) = \tau(m, n) + \tau(\tilde{m}, n)$,

2. $\tau(m, n + \tilde{n}) = \tau(m, n) + \tau(m, \tilde{n})$, and

3. $\tau(m\alpha\beta, n) = \tau(m, \alpha\beta n)$.

Let M_K and ${}_K N$ be ternary semimodules over a ternary semifield K. The map $\mu : M \times N \rightarrow M \otimes {}_K N$ defined by $\mu(m, n) = m \otimes n$ is called the *canonical middle linear map*. The function $\pi : \mathcal{FA}(M \times N) \rightarrow \mathcal{FA}(M \times N)/L$ defined by $\pi(x) = x + L$ for all $x \in \mathcal{FA}(M \times N)$ is an epimorphism of groups, which is called the *canonical projection*.

Lemma 1.5 ([9], p. 43): Let A and B be additive abelian groups. If $f : A \rightarrow B$ is a group homomorphism, and C is a subgroup of $\ker f$; then, there is a unique group homomorphism $\hat{f} : A/C \rightarrow B$ such that $\hat{f}(a + C) = f(a)$ for all $a \in A$, $\operatorname{im} \hat{f} = \operatorname{im} f$ and $\ker \hat{f} = \ker f / C$. Moreover, $\hat{f}$ is a group isomorphism if and only if f is a group epimorphism, and $C = \ker f$. In particular, $A/\ker f \cong \operatorname{im} f$.

By making use of Lemma 1.5, we get the following theorem.

Theorem 1.6. Let M_K and ${}_K N$ be ternary semimodules over a ternary semifield K and $\mu : M \times N \rightarrow M \otimes {}_K N$ be the canonical middle linear map. For any additive abelian group U over K, and any middle linear map $\psi : M \times N \rightarrow U$, there exists a unique group homomorphism $\tilde{\psi} : M \otimes {}_K N \rightarrow U$ such that $\psi = \tilde{\psi} \circ \mu$, that is, the following diagram commutes.

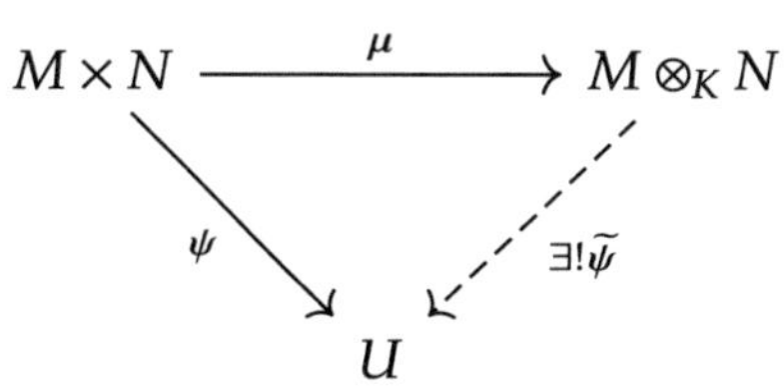

Proof: Let $(U, +)$ be an abelian group. By Theorem 1.1 ([9], p. 71) there exists a unique group homomorphism $\hat{\psi} : \mathcal{FA}(M \times N) \rightarrow U$ such that $\psi = \hat{\psi} \circ \varphi$.

Let L be the subgroup defined in Definition 1.3. Then, L is a subgroup of $\ker \hat{\psi}$ because ψ is a middle linear map and $\psi = \hat{\psi} \circ \varphi$.

Let $\pi : \mathcal{FA}(M \times N) \rightarrow \mathcal{FA}(M \times N)/L$ be the canonical projection. Since L is a subgroup of $\ker \hat{\psi}$, by Lemma 1.5, there exists a unique group homomorphism $\hat{\psi} : \mathcal{FA}(M \times N)/L \rightarrow U$ such that $\hat{\psi} = \tilde{\psi} \circ \pi$. Now, we obtain a group homomorphism $\tilde{\psi} : M \otimes {}_K N \rightarrow U$. Next, we will consider the diagram.

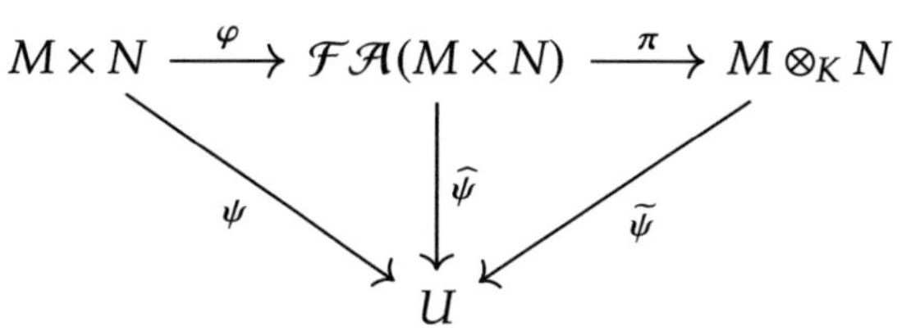

For each $m \in M$ and $n \in N$, the canonical middle linear map $\mu : M \times N \to M \otimes_K N$ satisfies $\mu(m,n) = m \otimes n = \varphi(m,n) + L = \pi(\varphi(m,n)) = \pi \circ \varphi(m,n)$. That is, $\mu = \pi \circ \varphi$. Hence, $\tilde{\psi} \circ \mu = \tilde{\psi} \circ (\pi \circ \varphi) = (\tilde{\psi} \circ \pi) \circ \varphi = \hat{\psi} \circ \varphi = \psi$. Last, let $\rho : M \otimes_K N \to U$ be a group homomorphism such that $\psi = \rho \circ \mu$, and let $\theta = \rho \circ \pi$. Consider the following diagram.

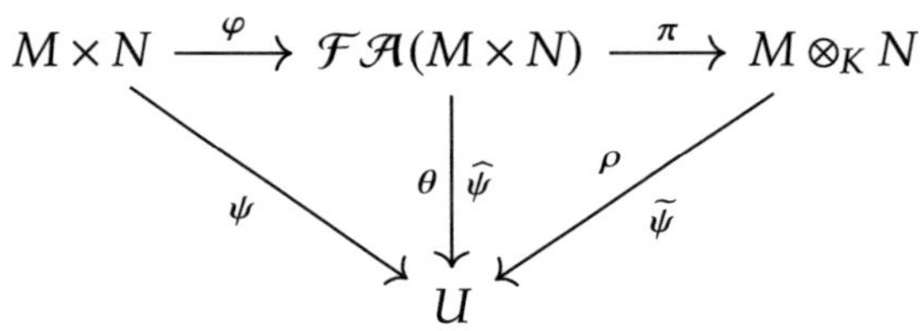

We have that $\theta \circ \varphi = (\rho \circ \pi) \circ \varphi = \rho \circ (\pi \circ \varphi) = \rho \circ \mu = \psi$. So, $\theta = \hat{\psi}$ because of the uniqueness of $\hat{\psi}$. Moreover, $\rho \circ \pi = \theta = \hat{\psi} = \tilde{\psi} \circ \pi$. By the uniqueness of $\tilde{\psi}$, we obtain that $\rho = \tilde{\psi}$.

Example 1.7

1. Let $u = [a_1, a_2, \ldots, a_m] \in \mathbb{R}^m$ and $v = [b_1, b_2, \ldots, b_n] \in \mathbb{R}^n$. Define a function $\psi : \mathbb{R}^m \times \mathbb{R}^n \to \mathcal{M}_{m \times n}(\mathbb{R})$ given by

$$\psi(u,v) = u^t v = \begin{bmatrix} a_1 \\ a_2 \\ \vdots \\ a_m \end{bmatrix} \begin{bmatrix} b_1 & b_2 & \cdots & b_n \end{bmatrix} = \begin{bmatrix} a_1b_1 & a_1b_2 & \cdots & a_1b_n \\ a_2b_1 & a_2b_2 & \cdots & a_2b_n \\ \vdots & \vdots & \ddots & \vdots \\ a_mb_1 & a_mb_2 & \cdots & a_mb_n \end{bmatrix}. \quad (8)$$

We can see that ψ is a middle linear map. By Theorem 1.6, there exists a unique group homomorphism $\tilde{\psi} : \mathbb{R}^m \otimes_{\mathbb{R}} \mathbb{R}^n \to \mathcal{M}_{m \times n}(\mathbb{R})$ such that $\psi(u,v) = \tilde{\psi}(u \otimes v) = u^t v$.

2. Let $A = [a_{ij}]_{n \times n} \in \mathcal{M}_n(\mathbb{R})$ and $B = [b_{ij}]_{n \times n} \in \mathcal{M}_n(\mathbb{R})$. Define a function $\psi : \mathcal{M}_n(\mathbb{R}) \times \mathcal{M}_n(\mathbb{R}) \to \mathcal{M}_{n^2}(\mathbb{R})$ given by

$$\psi(A,B) = [a_{ij}B] = \begin{bmatrix} a_{11}B & a_{12}B & \cdots & a_{1n}B \\ a_{21}B & a_{22}B & \cdots & a_{2n}B \\ \vdots & \vdots & \ddots & \vdots \\ a_{n1}B & a_{n2}B & \cdots & a_{nn}B \end{bmatrix}. \quad (9)$$

We can see that ψ is a middle linear map. By Theorem 1.6, there exists a unique group homomorphism $\tilde{\psi} : \mathcal{M}_n(\mathbb{R}) \otimes_{\mathbb{R}} \mathcal{M}_n(\mathbb{R}) \to \mathcal{M}_{n^2}(\mathbb{R})$ such that $\psi(A,B) = \tilde{\psi}(A \otimes B) = [a_{ij}B]$.

Let M_K, ${}_KN$, and ${}_KU$ be ternary semimodules over a ternary semifield K. A *bilinear map* over K from $M \times N$ to U is a function $T : M \times N \to U$ such that for all $m, \tilde{m} \in M$, $n, \tilde{n} \in N$ and $\alpha, \beta \in K$,

1. $T(m + \tilde{m}, n) = T(m,n) + T(\tilde{m}, n)$,

2. $T(m, n + \tilde{n}) = T(m,n) + T(m, \tilde{n})$, and

3. $T(m\alpha\beta, n) = \alpha\beta T(m,n) = T(m, \alpha\beta n)$.

The map $\mu : M \times N \to M \otimes_K N$ is given by $\mu(m,n) = m \otimes n$ and is called the *canonical bilinear map* (over a ternary semifield K).

By Theorem 1.6, the following result is derived.

Corollary 1.8 Let ${}_K M_K$, ${}_K N$, and ${}_K U$ be ternary semimodules over a ternary semifield K, and $\mu : M \times N \to M \otimes_K N$ the canonical bilinear map. For any bilinear map $\psi : M \times N \to U$, there exists a unique K-ternary semimodule homomorphism $\tilde{\psi} : M \otimes_K N \to U$ such that $\psi = \tilde{\psi} \circ \mu$, that is, the following diagram commutes.

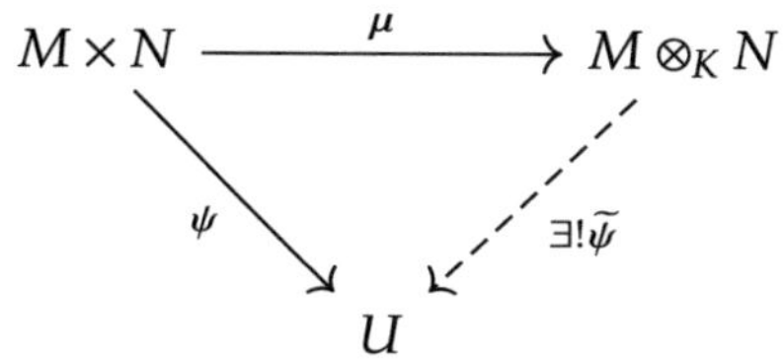

Example 1.9 We will apply Corollary 1.8 to show that $\mathbb{Q}_0^- \otimes_{\mathbb{Q}_0^-} F[0,1] \cong F[0,1]$.

Let $\mu : \mathbb{Q}_0^- \times F[0,1] \to \mathbb{Q}_0^- \otimes_{\mathbb{Q}_0^-} F[0,1]$ be a canonical bilinear map. That is, $\mu(a,f) = a \otimes f$, for all $a \in \mathbb{Q}_0^-$ and $f \in F[0,1]$.

Define $\psi : \mathbb{Q}_0^- \times F[0,1] \to F[0,1]$ by $\psi(a,f) = (-1)af$, where $a \in \mathbb{Q}_0^-$ and $f \in F[0,1]$. We can show that ψ is a bilinear map.

By the universal mapping property, there exists a unique ternary homomorphism $\tilde{\psi} : \mathbb{Q}_0^- \otimes_{\mathbb{Q}_0^-} F[0,1] \to F[0,1]$ such that $\psi = \tilde{\psi} \circ \mu$. That is, $\tilde{\psi}(a \otimes f) = \tilde{\psi}(\mu(a,f)) = \psi(a,f) = (-1)af$ for all $a \in \mathbb{Q}_0^-$ and $f \in F[0,1]$.

We will show that $\tilde{\psi}$ is an isomorphism. Let $f \in F[0,1]$. We have $-1 \otimes f \in \mathbb{Q}_0^- \otimes_{\mathbb{Q}_0^-} F[0,1]$ and $\tilde{\psi}(-1 \otimes f) = (-1)(-1)f = f$. So, $\tilde{\psi}$ is surjective.

Next, we will show that $\ker\tilde{\psi} = \{0\}$. Let $a \otimes f \in \mathbb{Q}_0^- \otimes_{\mathbb{Q}_0^-} F[0,1]$. Consider $a \otimes f = (-1)(-1)a \otimes f = -1 \otimes (-1)af$. Since $F[0,1]$ is a left $\mathbb{Q}_0^-$-ternary semimodule, $a \otimes f = -1 \otimes g$ for some $g \in F[0,1]$.

Let $-1 \otimes g \in \ker\tilde{\psi}$. Then, $\tilde{\psi}(-1 \otimes g) = 0$. Thus, $g = (-1)(-1)g = 0$, that is, $a \otimes f = -1 \otimes 0 = 0$. Hence, $\ker\tilde{\psi} = \{0\}$, so $\tilde{\psi}$ is injective. Consequently, $\tilde{\psi}$ is an isomorphism.

By Theorem 1.6, the following results about the tensor products of two ternary semimodule homomorphisms are derived.

Proposition 1.10 Let M_K, M'_K, ${}_K N$, and ${}_K N'$ be ternary semimodules over a ternary semifield K. If $f : M \to M'$ and $g : N \to N'$ are right and left K-ternary semimodule homomorphisms, respectively; then, there exists a unique group homomorphism h from $M \otimes_K N$ into $M' \otimes_K N'$ such that for all $m \in M$ and $n \in N$, $h(m \otimes n) = f(m) \otimes g(n)$.

The unique group homomorphism h in Proposition 1.10 is denoted by $f \otimes g : M \otimes_K N \to M' \otimes_K N'$.

Lemma 1.11 Let K and S be ternary semifields, and ${}_S M_K$ and ${}_K N$ be ternary semimodules as indicated. Then, $M \otimes_K N$ is a left ternary semimodule over S defined by $s_1 s_2 (m \otimes n) = s_1 s_2 m \otimes n$ for all $s_1, s_2 \in S$, $m \in M$, and $n \in N$.

Theorem 1.12 Let K and S be ternary semifields, and ${}_S M_K$, ${}_S M'_K$, ${}_K N$, and ${}_K N'$ be ternary semimodules as indicated. If $f : M \to M'$ is a right K-ternary semimodule homomorphism, and $g : N \to N'$ is a left K-ternary semimodule homomorphism; then, $f \otimes g$ is a left S-ternary semimodule homomorphism.

Proof: Let $f : M \to M'$ be an (S,K)-ternary bisemimodule homomorphism and $g : N \to N'$ be a left K-ternary semimodule homomorphism. By Lemma 1.11, we obtain that $M \otimes_K N$ and $M' \otimes_K N'$ are left S-ternary semimodules. Moreover, there exists a unique group homomorphism $f \otimes g : M \otimes_K N \to M' \otimes_K N'$ such that $(f \otimes g)(m \otimes n) = f(m) \otimes g(n)$ for all $m \in M$ and $n \in N$. By Lemma 1.11, we obtain that $s_1 s_1 \left(\sum_{i=1}^{p} m_i \otimes n_i \right) = \sum_{i=1}^{p} s_1 s_2 m_i \otimes n_i$ for all $s_1, s_2 \in S$ and $\sum_{i=1}^{p} m_i \otimes n_i \in M \otimes_K N$ then

$$(f \otimes g)\left(s_1 s_1 \left(\sum_{i=1}^{p} m_i \otimes n_i \right) \right) = s_1 s_2 \left((f \otimes g)\left(\sum_{i=1}^{p} m_i \otimes n_i \right) \right). \tag{10}$$

Therefore, $f \otimes g$ is a left S-ternary semimodule homomorphism.

Proposition 1.13 Let K be a ternary semifield, and M_K, M'_K, M''_K, ${}_K N$, ${}_K N'$ and ${}_K N''$ be ternary semimodules. If $f : M \to M'$ and $f' : M' \to M''$ are right K-ternary semimodule homomorphisms, $g : N \to N'$ and $g' : N' \to N''$ are left K-ternary semimodule homomorphisms. Then,

$$(f' \otimes g') \circ (f \otimes g) = (f' \circ f) \otimes (g' \circ g) : M \otimes_K N \to M'' \otimes_K N'' \tag{11}$$

is a group homomorphism. If f and g are right and left K-ternary semimodule isomorphisms, respectively; then, $f \otimes g$ is a group isomorphism and $(f \otimes g)^{-1} = f^{-1} \otimes g^{-1}$ is also a group isomorphism.

Proof: We can see that $\left(\tilde{f} \otimes \tilde{g}\right) \circ (f \otimes g)$ and $\left(\tilde{f} \circ f\right) \otimes (\tilde{g} \circ g)$ are K-ternary semimodule homomorphisms. It suffices to show that equality holds. Let $m \in M$ and $n \in N$. Then,

$$\begin{aligned} \left(\tilde{f} \otimes \tilde{g}\right) \circ (f \otimes g)(m \otimes n) &= \left(\tilde{f} \otimes g\right)(f(m) \otimes g(n)) \\ &= \left(\tilde{f} \circ f\right)(m) \otimes (\tilde{g} \circ g)(n) \\ &= \left(\tilde{f} \circ f\right) \otimes (\tilde{g} \circ g)(m \otimes n) \end{aligned} \tag{12}$$

Hence, $(f' \otimes g') \circ (f \otimes g) = (f' \circ f) \otimes (g' \circ g)$. For the last statement, it is straightforward.

4. The n-fold tensor products

Definition 1.14 Let K be a ternary semifield, and let $M_1, M_2, \dots, M_n$ and N be appropriate K-ternary semimodules. An n-multilinear map or simply multilinear map (over K) from $M_1 \times M_2 \times \cdots \times M_n$ into N is a function $f : M_1 \times M_2 \times \cdots \times M_n \to N$ such that for all $\alpha, \beta \in K$, $m_i \in M_i$ and $m'_j \in M_j$ for all $i, j = 1, 2, \dots n$,

1. $f\left(m_1, \dots, m_{j-1}, m_j + m'_j, m_{j+1}, \dots, m_n\right) = f(m_1, \dots, m_{j-1}, m_j, m_{j+1}, \dots, m_n) + f\left(m_1, \dots, m_{j-1}, m'_j, m_{j+1}, \dots, m_n\right)$, and

2. $f(m_1, \dots, m_{j-1}, \alpha\beta m_j, m_{j+1}, \dots, m_n) = \alpha\beta f(m_1, \dots, m_{j-1}, m_j, m_{j+1}, \dots, m_n)$

We can generalize the tensor products of two ternary semimodules over ternary semifields to tensor products of ternary semimodules $M_1, M_2, \dots, M_n$ over the same ternary semifield K, which we simply call *the n-fold tensor product* and write this as $M_1 \otimes_K M_2 \otimes_K \cdots \otimes_K M_n$ by dropping all parentheses.

For example, let $n \geq 2$ and $M_1, M_2, \ldots, M_n$ be ternary semimodules over a ternary semifield K. Then, the function

$$\mu_n : M_1 \times M_2 \times \cdots \times M_n \to M_1 \otimes_K M_2 \otimes_K \cdots \otimes_K M_n$$

defined by $(m_1, m_2, \ldots, m_n) \mapsto m_1 \otimes m_2 \otimes \cdots \otimes m_n$ is a multilinear map. We call the function μ_n a *canonical multilinear map*.

Theorem 1.15: (The universal mapping property of the n-fold tensor product) Let $n \geq 2$, and $M_1, M_2, \ldots, M_n$ and N be appropriate ternary semimodule over the same ternary semifield K. For any multilinear map, $\psi : M_1 \times M_2 \times \cdots \times M_n \to N$, there exists a unique K-ternary semimodule homomorphism $\tilde{\psi} : M_1 \otimes_K M_2 \otimes_K \cdots \otimes_K M_n \to N$ such that $\psi = \tilde{\psi} \circ \mu_n$, i.e., the following diagram commutes.

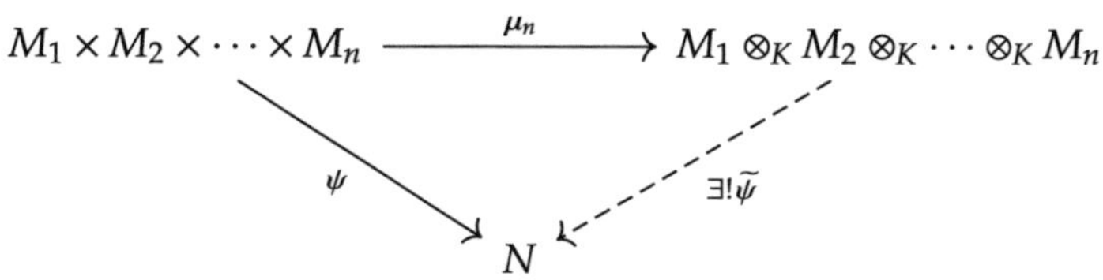

Proof: Let $\psi : M_1 \times M_2 \times \cdots \times M_n \times M_{n+1} \to N$ be a multilinear map. For each $m \in M_{n+1}$, let $\psi_m : M_1 \times M_2 \times \cdots \times M_n \to N$ be defined by

$$\psi_m(m_1, m_2, \ldots, m_n) = \psi(m_1, m_2, \ldots, m_n, m) \text{ for all } m_i \in M_i, i = 1,2,3, \ldots, n. \quad (13)$$

We can see that ψ_m is a multilinear map. By the induction hypothesis, there exists a unique K-ternary semimodule homomorphism $\tilde{\psi}_m : M_1 \otimes_K M_2 \otimes_K \cdots \otimes_K M_n \to N$ such that $\psi_m = \tilde{\psi}_m \circ \mu_n$.

Let $\mathcal{B} : (M_1 \otimes_K M_2 \otimes_K \cdots \otimes_K M_n) \times M_{n+1} \to N$ be defined by $\mathcal{B}(x, m) = \tilde{\psi}_m(x)$ for all $x \times M_1 \otimes_K M_2 \otimes_K \circ \otimes_K M_n$ and $m \in M_{n+1}$. Since $\tilde{\psi}_m$ is unique, $\mathcal{B}$ is well-defined. Next, we show that $\mathcal{B}$ is a bilinear map.

Let $\alpha, \beta \in K$, $x, y \in M_1 \otimes_K M_2 \otimes_K \circ \otimes_K M_n$ and $s, t \in M_{n+1}$. Then,

$$\mathcal{B}(x + y, s) = \tilde{\psi}_s(x + y) = \tilde{\psi}_s(x) + \tilde{\psi}_s(y) = \mathcal{B}(x, s) + \mathcal{B}(y, s), \quad (14)$$

and

$$\mathcal{B}(\alpha\beta x, s) = \tilde{\psi}_s(\alpha\beta x) = \alpha\beta\tilde{\psi}_s(x) = \alpha\beta\mathcal{B}(x, s). \quad (15)$$

Moreover, we have to show that $\mathcal{B}(x, s + t) = \mathcal{B}(x, s) + \mathcal{B}(x, t)$ and $\mathcal{B}(x, \alpha\beta s) = \alpha\beta\mathcal{B}(x, s)$. That is, $\tilde{\psi}_{s+t} = \tilde{\psi}_s + \tilde{\psi}_t$ and $\tilde{\psi}_{\alpha\beta s} = \alpha\beta\tilde{\psi}_s$, respectively. Because $\tilde{\psi}_{s+t}$ is a unique K-ternary semimodule homomorphism such that $\psi_{s+t} = \tilde{\psi}_{s+t} \circ \mu_n$; therefore, it suffices to show that $(\tilde{\psi}_s + \tilde{\psi}_t) \circ \mu_n = \psi_{s+t}$. Let $m_i \in M_i$ for all $i = 1, 2, \ldots, n$. Then,

$$\begin{aligned}
(\tilde{\psi}_s + \tilde{\psi}_t) \circ \mu_n(m_1, m_2, \ldots, m_n) &= (\tilde{\psi}_s \circ \mu_n)(m_1, m_2, \ldots, m_n) + (\tilde{\psi}_t \circ \mu_n)(m_1, m_2, \ldots, m_n) \\
&= \psi_s(m_1, m_2, \ldots, m_n) + \psi_t(m_1, m_2, \ldots, m_n) \\
&= \psi(m_1, m_2, \ldots, m_n, s) + \psi(m_1, m_2, \ldots, m_n, t) \quad (16) \\
&= \psi(m_1, m_2, \ldots, m_n, s + t) \\
&= \psi_{s+t}(m_1, m_2, \ldots, m_n).
\end{aligned}$$

Similarly, $\tilde{\psi}_{\alpha\beta s}$ is a unique K-ternary semimodule homomorphism such that $\psi_{\alpha\beta s} = \tilde{\psi}_{\alpha\beta s} \circ \mu_n$. So, we will show that $\alpha\beta\tilde{\psi}_s \circ \mu_n = \psi_{\alpha\beta s}$. Let $m_i \in M_i$ for all $i = 1, 2, \ldots, n$. Then,

$$\begin{aligned}((\alpha\beta\tilde{\psi}_s) \circ \mu_n)(m_1, m_2, \ldots, m_n) &= \alpha\beta\tilde{\psi}_s \circ \mu_n(m1, m2, \ldots, m_n) \\ &= \alpha\beta\psi_s(m_1, m_2, \ldots, m_n) \\ &= \alpha\beta\psi(m_1, m_2, \ldots, m_n, s) \\ &= \psi(m_1, m_2, \ldots, m_n, \alpha\beta s) \\ &= \psi_{\alpha\beta s}(m_1, m_2, \ldots, m_n).\end{aligned} \tag{17}$$

Hence, $\mathcal{B}$ is a bilinear map.

For each $m_i \in M_i$ for all $i = 1, 2, \ldots, n, n+1$, we consider the bilinear map $\mu : (M_1 \otimes_K M_2 \otimes_K \cdots \otimes_K M_n) \times M_{n+1} \to (M_1 \otimes_K M_2 \otimes_K \cdots \otimes_K M_n) \otimes_K M_{n+1}$. We can see that

$$\begin{aligned}\mu(\mu_n(m_1, m_2, \ldots, m_n), m_{n+1}) &= \mu_n(m_1, m_2, \ldots, m_n) \otimes m_{n+1} \\ &= (m_1 \otimes m_2 \otimes \cdots \otimes m_n) \otimes m_{n+1} \\ &= m_1 \otimes m_2 \otimes \cdots \otimes m_n \otimes m_{n+1} \\ &= \mu_{n+1}(m_1, m_2, \ldots, m_n, m_{n+1}).\end{aligned} \tag{18}$$

Fix $i \in \{1,2,3, \ldots, n, n+1\}$ and let $\alpha, \beta \in K$, $m_i, m_i' \in M_i$ for all i. If $i = 1, 2, \ldots, n$. Then,

$$\begin{aligned}\mu_{n+1}(m_1, m_2, \ldots, m_i + m_i', \ldots, m_n, m_{n+1}) &= \mu(\mu_n(m_1, m_2, \ldots, m_i + m_i', \ldots, m_n), m_{n+1}) \\ &= \mu(\mu_n(m_1, m_2, \ldots, m_i, \ldots, m_n) + \mu_n(m_1, m_2, \ldots, m_i', \ldots, m_n), m_{n+1}) \\ &= \mu(\mu_n(m_1, m_2, \ldots, m_i, \ldots, m_n), m_{n+1}) + \mu(\mu_n(m_1, m_2, \ldots, m_i', \ldots, m_n), m_{n+1}) \\ &= \mu_{n+1}(m_1, m_2, \ldots, m_i, \ldots, m_n, m_{n+1}) + \mu_{n+1}(m_1, m_2, \ldots, m_i', \ldots, m_n, m_{n+1}),\end{aligned} \tag{19}$$

and

$$\begin{aligned}\mu_{n+1}(m_1, m_2, \ldots, \alpha\beta m_i, \ldots, m_n, m_{n+1}) &= \mu(\mu_n(m_1, m_2, \ldots, \alpha\beta m_i, \ldots, m_n), m_{n+1}) \\ &= \mu(\alpha\beta\mu_n(m_1, m_2, \ldots, m_i, \ldots, m_n), m_{n+1}) \\ &= \alpha\beta\mu(\mu_n(m_1, m_2, \ldots, m_i, \ldots, m_n), m_{n+1}) \\ &= \alpha\beta\mu_{n+1}(m_1, m_2, \ldots, m_i, \ldots, m_n, m_{n+1}).\end{aligned} \tag{20}$$

If $i = n + 1$. Then,

$$\begin{aligned}\mu_{n+1}(m_1, m_2, \ldots, m_n, m_{n+1} + m_{n+1}') &= \mu(\mu_n(m_1, m_2, \ldots, m_n), m_{n+1} + m_{n+1}') \\ &= \mu(\mu_n(m_1, m_2, \ldots, m_n), m_{n+1}) + \mu(\mu_n(m_1, m_2, \ldots, m_n), m_{n+1}') \\ &= \mu_n + 1(m_1, m_2, \ldots, m_n, m_{n+1}) + \mu_n + 1(m_1, m_2, \ldots, m_n, m_{n+1}'),\end{aligned} \tag{21}$$

and

$$\begin{aligned}\mu_{n+1}(m_1, m_2, \ldots, m_n, \alpha\beta m_{n+1}) &= \mu(\mu_n(m_1, m_2, \ldots, m_n), \alpha\beta m_{n+1}) \\ &= \alpha\beta\mu(\mu_n(m_1, m_2, \ldots, m_n), m_{n+1}) \\ &= \alpha\beta\mu_{n+1}(m_1, m_2, \ldots, m_n, m_{n+1}).\end{aligned} \tag{22}$$

Hence, μ_{n+1} is an $(n+1)$-multilinear map.

Next, we consider the case $n+1$, where $\psi : M_1 \times M_2 \times \cdots \times M_{n+1} \to N$ is a multilinear map. Because $\mathcal{B}$ and μ are bilinear maps and by the induction hypothesis, so there exists a unique K-ternary module homomorphism $\tilde{\psi}$: $M1 \otimes_K M2 \otimes_K \cdots \otimes_K M_n \otimes_K M_{n+1} \to N$ such that $\mathcal{B} = \tilde{\psi} \circ \mu$.

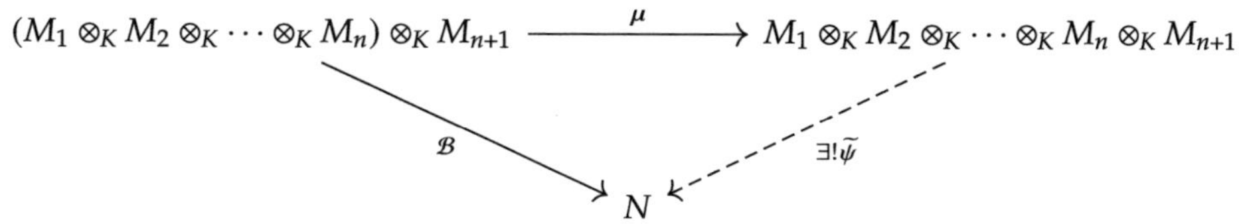

Consider the following diagram.

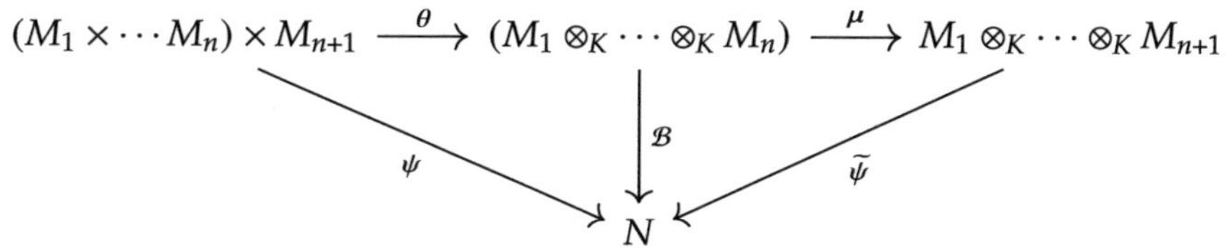

where θ is a function from $(M_1 \times M_2 \times \cdots \times M_n) \times M_{n+1}$ into $(M_1 \otimes_K M_2 \otimes_K \cdots \otimes_K M_n) \times M_{n+1}$ defined by $\theta((m_1, m_2, \ldots, m_n), m_{n+1}) = (\mu_n(m_1, m_2, \ldots, m_n), m_{n+1})$. To show $\psi = \tilde{\psi} \circ \mu_{n+1}$, let $m_i \in M_i$ for all $i = 1,2,3, \ldots, n+1$. Then,

$$\begin{aligned} \tilde{\psi} \circ \mu_{n+1}(m_1, m_2, \ldots, m_n, m_{n+1}) &= \tilde{\psi}(\mu(\mu_n(m_1, m_2, \ldots, m_n), m_{n+1})) \\ &= \mathcal{B}(\mu_n(m_1, m_2, \ldots, m_n), m_{n+1}) \\ &= \tilde{\psi}_{m_{n+1}}(\mu_n(m_1, m_2, \ldots, m_n)) \\ &= \psi_{m_{n+1}}(m_1, m_2, \ldots, m_n) \\ &= \psi(m_1, m_2, \ldots, m_n, m_{n+1}). \end{aligned} \tag{23}$$

Thus, $\psi = \tilde{\psi} \circ \mu_{n+1}$.

Finally, we show that $\tilde{\psi}$ is unique. Suppose that κ : $M_1 \otimes_K M_2 \otimes_K \cdots \otimes_K M_n \otimes_K M_{n+1} \to N$ is a K-ternary semimodule homomorphism such that $\psi = \kappa \circ \mu_{n+1}$. For each $s \in M_{n+1}$, let $\kappa_s : M_1 \otimes_K M_2 \otimes_K \cdots \otimes_K M_n \to N$ be defined by $\kappa_s(x) = \kappa(\mu(x, s))$ for all $x \in M_1 \otimes_K M_2 \otimes_K \cdots \otimes_K M_n$. Then, κ_s is a K-ternary semimodule homomorphism. Let $m_i \in M_i$ for all $i = 1,2,3, \ldots, n$. Then,

$$\begin{aligned} (\kappa_s \circ \mu_n)(m_1, m_2, \ldots, m_n) &= \kappa(\mu(\mu_n(m_1, m_2, \ldots, m_n), s)) \\ &= \kappa(\mu_{n+1}(m_1, m_2, \ldots, m_n, s)) \\ &= \psi(m_1, m_2, \ldots, m_n, s) \\ &= \psi_s(m_1, m_2, \ldots, m_n). \end{aligned} \tag{24}$$

Thus, $\kappa_s \circ \mu_n = \psi_s$. By the uniqueness of $\tilde{\psi}_s$, we obtain that $\kappa_s = \tilde{\psi}_s$. Now, we show that $\mathcal{B} = \kappa \circ \mu$. Let $x \in M_1 \otimes_K M_2 \otimes_K \cdots \otimes_K M_n$ and $s \in M_{n+1}$. Then, $(\kappa \circ \mu)(x, s) = \kappa(\mu(x, s)) = \kappa_s(x) = \tilde{\psi}_s(x) = \mathcal{B}(x, s) = \tilde{\psi} \circ \mu(x, s)$. By the uniqueness of $\tilde{\psi}$, we obtain that $\kappa = \tilde{\psi}$. Hence, $\tilde{\psi}$ is unique.

5. Exact sequences

For the last section, we investigate short exact sequences of ternary semimodules and flat ternary semimodules. By the way, we follow the definitions of short exact sequences and flat semimodules from Refs. [6, 10] but we change the semimodules to the ternary semimodules.

Theorem 1.16 Let K be a ternary semifield, M be a left K-ternary semimodule and $\{V_i \mid i \in I\}$ be a family of right K-ternary semimodules. If each V_i for all $i \in I$ is M-flat; then, the direct sum V_i is M-flat.

Proof: Let id_M be the identity function on a K-ternary semimodule M. Consider the following diagram

$$\begin{array}{ccccc} V_i \otimes_K N & & & & \left(\bigoplus_{i\in I} V_i\right) \otimes_K N \\ \Big\downarrow \mathrm{id}_{V_i} \otimes \mathrm{id}_N & & & & \Big\downarrow \mathrm{id}_{\bigoplus_{i\in I} V_i} \otimes \mathrm{id}_N \\ V_i \otimes_K M & \underset{\pi_i}{\overset{\iota_i}{\rightleftarrows}} & \bigoplus_{i\in I} (V_i \otimes_K M) & \underset{\psi}{\overset{\psi^{-1}}{\rightleftarrows}} & \left(\bigoplus_{i\in I} V_i\right) \otimes_K M \end{array} \qquad (*)$$

where $\pi_i : \oplus_{i\in I}(V_i \otimes_K M) \to V_i \otimes_K M$ given by $\pi_i(\{v_i \otimes m\}) = v_i \otimes m$ for all $i \in I$ and $\iota_i : V_i \otimes_K M \to \oplus_{i\in I}(V_i \otimes_K M)$ given by $\iota_i(v_i \otimes m) = \{v_j \otimes m\}$ where

$$v_j \otimes m = \begin{cases} v_i \otimes m & \text{if} \quad i = j, \\ 0 & \text{if} \quad i \neq j \end{cases} \tag{25}$$

for all $i, j \in I$. There exists a group isomorphism $\psi : (\oplus_{i\in I} V_i) \otimes_K M \to \oplus_{i\in I}(V_i \otimes_K M)$ such that $\psi(\{v_i\} \otimes m) = \{v_i \otimes m\}$.

Assume that V_i is M-flat for each $i \in I$. Then,

$$0 = (\mathrm{id}_{V_i} \otimes \mathrm{id}_N)(v_i \otimes n) = \mathrm{id}_{V_i}(v_i) \otimes \mathrm{id}_N(n) = v_i \otimes m \tag{26}$$

for some $m \in M$. By diagram (*), we have that

$$0 = \psi^{-1}(\iota_i(v_i \otimes m)) = \psi^{-1}(\{v_i \otimes m\}) = \{v_i\} \otimes m. \tag{27}$$

So, $(\mathrm{id}_{\oplus_{i\in I} V_i} \otimes \mathrm{id}_N)(\{v_i\} \otimes n) = \mathrm{id}_{\oplus_{i\in I} V_i}(\{v_i\}) \otimes \mathrm{id}_N(n) = \{v_i\} \otimes m = 0$.
That is, $\ker(\mathrm{id}_{\oplus_{i\in I} V_i} \otimes \mathrm{id}_N) = 0$. Hence, $\oplus_{i\in I} V_i$ is M-flat.

Acknowledgements

We thank the referees for their suggestions and comments on the manuscript.

Author details

Thorranin Pawaputanon Na Mahasarakham[1], Ruangvarin Intarawong Sararnrakskul[2]* and Nissara Sirasuntorn[2]

1 Piboonbumpen Demonstration School, Burapha University, Chonburi, Thailand

2 Department of Mathematics, Faculty of Science, Srinakharinwirot University, Bangkok, Thailand

*Address all correspondence to: ruangvarin@g.swu.ac.th

References

[1] Lehmer DH. A ternary analogue of abelian groups. The American Journal of Mathematics. 1932;**54**:329-338

[2] Los J. On the extending of models I. Fundamenta Mathematicae. 1955;**42**: 38-54

[3] Lister WG. Ternary rings. Transactions of the American Mathematical Society. 1971;**154**:37-55

[4] Intarawong R. Tensor Products of Modules over Semifields. Bangkok: Chulalongkorn University; 2003

[5] Dutta TK, Kar S. On ternary semifields. General Algebra and Applications. 2004;**24**:85-98

[6] Al-Thani HMJ. Flat semimodules. International Journal of Mathematics and Mathematical Sciences. 2004;**17**: 873-880

[7] Amyari M, Moslehian S. Approximate homomorphisms of ternary semigroups. Letters in Mathematical Physics. 2006; **77**:1-9

[8] Dummit DS, Foote RM. Abstract Algebra. New Jersey: John Wiley and Sons; 2004

[9] Hungerford TW. Algebra. New York: Springer-Verlag; 1974

[10] Abuhlail JY. Some remarks on tensor products and flatness of semimodules. Semigroup Forum. 2014;**88**:732-738

Chapter 2

On the Classical and Deformed Korteweg-de Vries Equation

Abderrahman El Boukili, Hicham Lekbich, Tahir Toghrai, Najim Mansour and Moulay Brahim Sedra

Abstract

Given the general nonlinear partial differential equations and the importance of the Korteweg-de Vries equation (KdV) in physics, this chapter presents a basic survey of the two-dimensional Korteweg-de Vries model. We begin by examining various symmetries of systems, and then explore the concept of integrability through two different methods: the Hamiltonian formalism and the existence of conserved quantities. By introducing the concept of q-deformation, we construct the corresponding q-deformation integrable model and the integrability of the resulting system is guaranteed by the existence of Lax pairs. We also present the KdV equation in the Moyal space of moments in its noncommutative version, we present the algebraic structure of the system and we study the integrability using the notion of Lax pair.

Keywords: Korteweg-de Vries equation, integrability, Lax generating technique, Q-deformation, Moyal momentum space

1. Introduction

The axis of nonlinear integrable systems is one of the interesting topics that has been recently examined from different points of view [1–3]. They are soluble models which present a very rich structure and are used in many fields of modern sciences such as mathematics, physics, chemistry and engineering sciences [3–5]. From a physical point of view, integrable systems play an important role in describing physical phenomena such as condensed matter, fluid dynamics, plasma physics, high energy physics and nonlinear optics [4, 6].

Nonlinear integrable systems involve nonlinear differential equations that can sometimes be solved exactly, but this is generally not a trivial task. However, we introduce some rigorous background such as the theory of pseudo-differential operators, Lie algebras, the inverse scattering method and some physical methods such as the related Lax formulation is needed [7–9]. The particularity of two-dimensional integrable systems lies in their revolutionary properties worthy of nonlinear differential equations such as Burgers, KP, Boussinesq, KdV [5].

In this work, we investigate some properties related to this prototype nonlinear differential equation, based on the classical Q-deformation version and the interesting system KdV of the Moyal momentum deformation, and show how to reinforce its central role [10, 11].

The composition of this work is as follows. Section 2 gives some interesting definitions of the 2d-KdV formula. Determining the Hamiltonian structure of the KdV equation, he guarantees the integrability of the KdV equation by the existence of infinitely many independent conserved quantities in the involution. in seconds In 3, we present some results related to the Lax representation of the q-deformed version. Using conventional notation and Q-analysis, we perform consistent algebraic computations based on pseudo-differential analysis and derive explicit loose pairing operators for KdV systems in the q-deformation framework. Section 4 presents some results related to his Lax representation of the Moyal $-\theta$ integrable hierarchy. Using the notation and analysis previously presented and developed in [12, 13], we perform coherent algebraic computations based on Moyal momentum analysis and integrable systems for some of the $\theta-$ deformation frames. We derive an explicit Lax pair operator for. Finally, Section 5 summarizes our research.

2. Classical 2d KdV equation

Note that the KdV equation is formulated without considering the integrability of this equation. It is a nonlinear dispersive partial differential equation for a function u of two real variables x and t, and its form is given by [2]:

$$\frac{\partial u}{\partial t} + au\frac{\partial u}{\partial x} + b\frac{\partial^3 u}{\partial x^3} = 0 \tag{1}$$

where $u(x,t)$ is a dynamic variable that plays the role of basin wave height.

The eq. (1) contains two arbitrary constants "a" and "b" which can be eliminated after a series of transformations namely:

- Transformation on spatial variables x (scale transformation):

$$x \to b^{1/3}x \tag{2}$$

- Transformation on the time variable t (time reversal):

$$t \to -t \tag{3}$$

- Transformation on the dynamic variable

$$u \to a^{-1}b^{1/3}u \tag{4}$$

After performing these transformations, the KdV equation can be reformulated in this way.

$$\frac{\partial u}{\partial t} = u\frac{\partial u}{\partial x} + \frac{\partial^3 u}{\partial x^3} \tag{5}$$

This equation is the most used in KdV theory. However to find the solutions of this equation and to classify them, we must know the symmetries of the equation. In fact, the KdV equation then has these four symmetries

$$\begin{aligned} &i)\ t \rightarrow d_1 + t, \\ &ii)\ x \rightarrow d_2 + x, \\ &iii)\ x \rightarrow \alpha x, t \rightarrow \alpha^3 t \text{ and } u \rightarrow \alpha^{-2} u \\ &iv)\ x \rightarrow x + vt, t \rightarrow t \text{ and } u \rightarrow u + v \end{aligned} \tag{6}$$

All these transformations are symmetries of the KdV equation. Note that transformation iii) is used for the classification of the conserved quantities of the system and that transformation iv) is considered as the Galilean invariance of such equations. However, the KdV equation does not have Lorentz invariance because of the separable treatment of the time and space variables.

2.1 Integrability through Hamiltonian structure

Let us consider the eq. (5), in order to give the Hamiltonian form of our equation, We need to calculate a a Poisson bracket structure and an Hamiltonian operator of the system [14]. In fact, we have to make, under somes assumptions, the KdV equation in the following Hamiltonian form:

$$\frac{\partial u}{\partial t} = \{u(x,t), H\} \tag{7}$$

Here $\{.,.\}$ represents the Poisson brackets and H is the Hamiltonian of our system. In fact, rewriting the KdV equation in the form

$$\begin{aligned} \frac{\partial u}{\partial t} &= \frac{\partial}{\partial x}\left(\frac{1}{2}u^2\right) + \frac{\partial}{\partial x}\left(\frac{\partial^2 u}{\partial x^2}\right) \\ &= \frac{\partial}{\partial x}\left(\frac{1}{2}u^2 + \frac{\partial^2 u}{\partial x^2}\right) \end{aligned} \tag{8}$$

we can then choose the following Hamiltonian:

$$H[u] = \int_{-\infty}^{+\infty} dx \left(\frac{1}{3!}u^3 - \frac{1}{2}\left(\frac{\partial u}{\partial x}\right)^2\right) \tag{9}$$

Based on the functional derivative defined by

$$\frac{\delta F(u(x))}{\delta u(x)} = \lim_{\varepsilon \to 0} \frac{1}{\varepsilon}\left(F[u(x) + \varepsilon\delta(x-y)] - F[u(x)]\right) \tag{10}$$

we can calculate the variation of H with respect to the dynamic variable u by

$$\frac{\delta H}{\delta u(x)} = \frac{1}{2}u^2 + \frac{\partial^2 u}{\partial x^2} \tag{11}$$

with this choice of H, the eq. (5) becomes

$$\frac{\partial u}{\partial t} = \frac{\partial}{\partial x}\frac{\delta H}{\delta u(x)} \tag{12}$$

by comparing the eqs. (7) and (12), we then find

$$\{u(x), H\} = \frac{\partial}{\partial x}\frac{\delta H}{\delta u(x)} \tag{13}$$

that is to say

$$\int_{-\infty}^{+\infty} dy \frac{\delta H}{\delta u(x)}\{u(x), u(y)\} = \frac{\partial}{\partial x}\frac{\delta H}{\delta u(x)} \tag{14}$$

in this case, the Poisson bracket of $u(x)$ and $u(y)$ is given by

$$\{u(x), u(y)\} = \frac{\partial}{\partial x}\delta(x-y) \tag{15}$$

This defines the Poisson structure associated to our KdV equation, however this bracket is antisymmetric and it verifies the Jacobi identity [2].

We can also define a second Hamiltonian structure of the KdV equation, for this we can choose a second structure of the Poisson bracket namely

$$\{u(x), u(y)\}_2 = \left(\frac{1}{3}\left(\frac{\partial}{\partial x}u(x) + u(x)\frac{\partial}{\partial x}\right) + \frac{\partial^3}{\partial x^3}\right)\delta(x-y) \tag{16}$$

and a second Hamiltonian operator

$$H_2[u] = \int_{-\infty}^{+\infty} dx \frac{1}{2}u^2(x) \tag{17}$$

with this choice, the equation of KdV becomes

$$\begin{aligned} \frac{\partial u}{\partial t} &= \{u(x,t), H_2\}_2 \\ &= \int_{-\infty}^{+\infty} dy \frac{\delta H_2}{\delta u(x)}\{u(x), u(y)\}_2 \\ &= u\frac{\partial u}{\partial x} + \frac{\partial^3 u}{\partial x^3} \end{aligned} \tag{18}$$

Remark

- This second Poison structure is also antisymmetric and it verifies the Jacobi identity, thus we can identify it with the Virasoro algebra with a specific charge [6].

- Based on the Lagrangian formalism, we can re-establish the KdV equation, indeed

$$\begin{aligned} L_{KdV} = &\frac{1}{2}\int_{-\infty}^{+\infty} dxdy u(x)\varepsilon(x-y)\frac{\partial u(y)}{\partial t} \\ &-\int_{-\infty}^{+\infty} dx\left(\frac{1}{3!}u^3(x) - \frac{1}{2}\left(\frac{\partial u(x)}{\partial x}\right)^2\right) \end{aligned} \tag{19}$$

The Euler Lagrange equations obtained are

$$\int_{-\infty}^{+\infty} dy\varepsilon(x-y)\frac{\partial u(y)}{\partial t} = \frac{1}{2}u^2(x) + \frac{\partial u^2(x)}{\partial x^2} \tag{20}$$

this means that

$$\frac{\partial u}{\partial t} = u\frac{\partial u}{\partial x} + \frac{\partial^3 u}{\partial x^3} \tag{21}$$

so we have recovered the eq. (5). The existence of the term $\varepsilon(x-y)$ in the Lagrangian implies the nonlocality of our system.

2.2 Integratability through the conserved quantities

In general, a quantity $Q[u]$ is said to be conserved if for a bracket $\{,\}$ and an Hamiltonian H, we have th following equation

$$\frac{dQ[u]}{dt} = \{Q[u], H\} = 0 \tag{22}$$

if moreover we take this expression

$$Q[u] = \int_{-\infty}^{+\infty} dx\rho[u(x,t)] \tag{23}$$

we have

$$\frac{dQ[u]}{dt} = \int_{-\infty}^{+\infty} dx\frac{\partial \rho[u(x,t)]}{\partial t} = 0 \tag{24}$$

this implies the existence of the continuity equation of the following form

$$\frac{\partial \rho[u(x,t)]}{\partial t} + \frac{\partial j[u(x,t)]}{\partial x} = 0 \tag{25}$$

In the following, we write the KdV eq. (5) as an equation of continuity (8) and compare it with the eq. (25), which then gives us the following equations

$$\begin{aligned} \rho_0[u(x,t)] &= u(x,t) \\ j_0 &= -\left(\frac{1}{2}u^2 + \frac{\partial^2 u}{\partial x^2}\right) \end{aligned} \tag{26}$$

consequently

$$Q_0 = H_0 = \int_{-\infty}^{+\infty} dx\rho_0[u(x,t)] = \int_{-\infty}^{+\infty} dx\, u(x,t) \tag{27}$$

is a constant of movement. In the following, we redefine ρ_0 and H_0 in this way

$$\begin{aligned} \rho_0[u(x,t)] &= 3u(x,t) \\ H_0 &= \int_{-\infty}^{+\infty} dx \rho_0[u(x,t)] = 3\int_{-\infty}^{+\infty} dx\, u(x,t) \end{aligned} \tag{28}$$

and the KdV equation becomes

$$\frac{\partial}{\partial t}\left(\frac{1}{2}u^2\right) = \frac{\partial}{\partial x}\left(\frac{1}{3}u^3 - \frac{1}{2}\left(\frac{\partial u}{\partial x}\right)^2 + u\frac{\partial^2 u}{\partial x^2}\right) \tag{29}$$

let us pose

$$\begin{aligned} \rho_1[u(x,t)] &= \frac{1}{2}u^2 \\ j_1 &= \frac{1}{3}u^3 - \frac{1}{2}\left(\frac{\partial u}{\partial x}\right)^2 + u\frac{\partial^2 u}{\partial x^2} \end{aligned} \tag{30}$$

we have a continuity equation, therefore the second constant of motion is obtained by

$$H_1 = \int_{-\infty}^{+\infty} dx\, \frac{1}{2}(u(x,t))^2 \tag{31}$$

In addition, note that the Hamiltonian of KdV equation is given from eq. (9) in the form

$$H_{KdV} = \int_{-\infty}^{+\infty} dx\left(\frac{1}{3!}u^3 - \frac{1}{2}\left(\frac{\partial u}{\partial x}\right)^2\right) \tag{32}$$

which define a constant of movement, namely

$$\frac{dH_{KdV}}{dt} = \{H_{KdV}, H_{KdV}\} = 0 \tag{33}$$

then we can identify the two Hamiltonians H_2 and H_{KdV}.

From the Poisson bracket defined by the eq. (15) we can also identify the Hamiltonian H_2 with the generators defining the translations in the time

$$\{u(x,t), H_2\} = \frac{\partial u(x,t)}{\partial t} \tag{34}$$

and H_1 with the generator defining the translations in the space

$$\{u(x,t), H_1\} = \frac{\partial u(x,t)}{\partial x} \tag{35}$$

These are the first conserved quantities obtained by a direct calculation. However, Miura had found 11 conserved quantities [7], while Kruskal postulated that there must be an infinite number of independent conserved quantities which are then obtained by a relation of recurrences [2, 7].

$$\rho_n = 3(-1)^n v_{2n} \tag{36}$$

and

$$H_n = \int_{-\infty}^{+\infty} dx\ \rho_n = 3(-1)^n \int_{-\infty}^{+\infty} dx\ v_{2n} \tag{37}$$

with

$$\begin{aligned} &v_0 = u, \\ &v_n + i\frac{\partial v_{n-1}}{\partial x} + \frac{1}{6}\sum_{m=0}^{n-2} v_{n-m-2} v_m = 0 \end{aligned} \tag{38}$$

we notice that the first 3 values of ρ_n and H_n correspond to the eqs. (28), (30), (32).

2.3 Integrability criterion

In the following, we ensure the integrability of the KdV equation by seeking the existence of an infinite number of independent conserved quantities which are in evolution [8]. Indeed, the conserved quantities H_0, H_1 and H_2 satisfy the functional relation

$$\left({D_x}^3 + \frac{1}{3}(D_x u + u D_x)\right)\frac{\delta H_{n-1}}{\delta u(x)} = D_x \frac{\delta H_n}{\delta u(x)} \tag{39}$$

We can generalize this equation for higher orders of n, by defining the following recurrence relation [2, 5, 7].

$$\begin{aligned} &H_{-1} = 0 \\ &\left({D_x}^3 + \frac{1}{3}(D_x u + u D_x)\right)\frac{\delta H_{n-1}}{\delta u(x)} = D_x \frac{\delta H_n}{\delta u(x)} \end{aligned} \tag{40}$$

Equation (40) is very important to say that the conserved quantities are in evolution. Indeed

$$\begin{aligned} \{H_n, H_m\}_1 &= \int_{-\infty}^{+\infty} dx \frac{\delta H_n}{\delta u(x)} D_x \frac{\delta H_m}{\delta u(x)} \\ &= \int_{-\infty}^{+\infty} dx \frac{\delta H_{n-1}}{\delta u(x)} D_x \frac{\delta H_{m+1}}{\delta u(x)} \\ &= \{H_{n-1}, H_{m+1}\}_1 = \ldots \\ &= \{H_{-1}, H_{m+n+1}\}_1 = 0 \end{aligned} \tag{41}$$

by an iteration relation we can show that

$$\{H_n, H_m\}_1 = \{H_m, H_n\}_1 \tag{42}$$

Therefore this relation affirms that our conserved quantities are independent and that they are also in evolution. We can also show that

$$\{H_n, H_m\}_2 = 0 \tag{43}$$

To conclude, according to Liouville's theorem [1–4, 7], the KdV equation is integrable.

3. Q-deformation of KdV eq.

3.1 Q-deformed formalism

To understand the q-deformation, we begin the section with a general framework, the $\alpha-$ derivative, defined in the form

$$\partial_\alpha(fg) = \alpha(f)d_\alpha(g) + d_\alpha(f)g, \tag{44}$$

with f and g are two polynomial functions in an indeterminate x and its inverse x^{-1} and the parameter α is a linear map. As an example of the derivation α, we present Jackson's q-differential operator ∂_q, defined by [9].

$$\partial_q(f) = \frac{f(qx) - f(x)}{(q-1)x} \tag{45}$$

in this formalism the eq. (44) becomes

$$\partial_q(fg) = \eta_q(f).\partial_q(g) + \partial_q(f).g, \tag{46}$$

where the operator η_q named **q-shift operator** which is given by

$$\eta_q(f(x)) = f(qx). \tag{47}$$

we can define the commutation relation as follows

$$[f, g] = f \circ g - g \circ f, \tag{48}$$

where $\circ$ is an internal composition law which is given by

$$\begin{aligned} \partial_q \circ f &= \eta_q(f)\partial_q + \partial_q f, \\ \partial_q^{-1} \circ f &= \sum\nolimits_{k \geq 0} (-1)^k q^{-k(k+1)/2} \eta_q^{-k-1}\left(\partial_q^k f\right)\partial_q^{-k-1} \end{aligned} \tag{49}$$

to obtain the second relation, we used the following unitarity relation

$$\partial_q^{-1} \circ \partial_q \circ f = \partial_q \circ \partial_q^{-1} \circ f = f. \tag{50}$$

where ∂_q^{-1} is the formal inverse q-derivative of the q-derivative ∂_q. Note that the shift operator η_q does not commute with ∂_q because we have

$$\begin{aligned} \partial_q\left(\eta_q(f)\right) &= q\eta_q\left(\partial_q f\right), \\ \partial_q\left(\eta_q^k(f)\right) &= q^k \eta_q^k\left(\partial_q f\right), \quad k \in \mathbb{Z}. \\ \partial_q^m\left(\eta_q^k(f)\right) &= q^{k+m}\eta_q^k\left(\partial_q^m f\right), \quad k, m \in \mathbb{Z}. \end{aligned} \tag{51}$$

However, we can generalize eq. (49) for all n as follows

$$\partial_q^n \circ f = \sum_{k\geq 0} \binom{n}{k}_q \eta_q^{n-k} \left(\partial_q^k f\right) \partial_q^{n-k}, \tag{52}$$

where the $q-$ binomials take the form [15].

$$\binom{n}{k}_q = \frac{(n)_q (n-1)_q \cdot \ldots (n-k+1)_q}{(1)_q (2)_q \ldots (k)_q}, \quad \binom{n}{0}_q = 1, \tag{53}$$

and the numbers $(n)_q$ called the q-numbers are given by

$$(n)_q = \frac{q^n - 1}{q - 1}, \tag{54}$$

Explicitly, we can write the eq. (49) for positive and negative degrees as follows

$$\begin{aligned} \partial_q \circ f &= (\partial_q f) + \eta_q(f)\partial_q, \\ \partial_q^2 \circ f &= \left(\partial_q^2 f\right) + (q+1)\eta_q(\partial_q f)\partial_q + \eta_q^2(f)\partial_q^2, \\ \partial_q^3 \circ f &= \left(\partial_q^3 f\right) + (q^2+q+1)\eta_q\left(\partial_q^2 f\right)\partial_q \\ &\quad + (q^2+q+1)\eta_q^2(\partial_q f)\partial_q^2 + \eta_q^3(f)\partial_q^3, \end{aligned} \tag{55}$$

and

$$\begin{aligned} \partial_q^{-1} \circ f &= \eta_q^{-1}(f)\partial_q^{-1} - q^{-1}\eta_q^{-2}(\partial_q f)\partial_q^{-2} + q^{-3}\eta_q^{-3}\left(\partial_q^2 f\right)\partial_q^{-3} \\ &\quad - q^{-6}\eta_q^{-4}\left(\partial_q^3 f\right)\partial_q^{-4} + \frac{1}{q^{10}}\eta_q^{-5}\left(\partial_q^4 f\right)\partial_q^{-5} + \cdots \\ &\quad + (-1)^k q^{-(1+2+3+\cdots+k)}\eta_q^{-k-1}\left(\partial_q^k f\right)\partial_q^{-k-1} + \cdots \end{aligned} \tag{56}$$

3.2 Integrability through Lax pair technique

The purpose of this section is to investigate the integrability of some nonlinear models in the q-deformed version based on the existence of Lax pairs. Such studies are compatible with those already established in the literature by assuming the classical limit $q = 1$ [8, 14].

The basic idea is to consider a nonlinear system with the Lax representation [10].

$$L, \partial_t - B]_q = 0, \tag{57}$$

where $\partial_t \equiv \frac{\partial}{\partial t}$ and $[f, g]_q = f \circ g - f \circ g$. The eq. (57) and the associated pair of operators (L, B) are called the Lax-q differential equation and the Lax pair operator, respectively.

In this technique, we must fix from the start the q-differential operator L and the q-deformation of the sl_n-KdV hierarchy equation is given by

$$\frac{\partial L}{\partial t_k} = \left[\left(L^{\frac{k}{2}}\right)_+, L\right]_q \tag{58}$$

the eq. (57) is then equivalent to the following equation

$$[L, \partial_t - B]_q \equiv \left[L, \partial_{t_k} - \left(L^{\frac{k}{2}}\right)_+\right]_q = 0 \tag{59}$$

here the operator B is the analogue of the operator $\left(L^{\frac{k}{2}}\right)_+$ describing a q-deformation operator of conformal spin k.

Our goal is to find an operator B that satisfies the eq. (57), to simplify the task, we will make constraints on the operator B called *Ansatz*, namely:

$$B = \partial_q^n \circ L^m + \tilde{B}, \tag{60}$$

where $\tilde{B}$ is another operator with the same equiangular weights as B. Second, using this approach alleviates the problem of finding operator $\tilde{B}$. Below we perform an application to the q-modified KdV equation.

3.3 Q-KdV equation

The discovery of Lax pairs is the first development major in modern integrable systems theory. By building a pair of Lax from one integrable Hamiltonian system, we obtain the solution of the system. Because the Lax equation is the equivalent of the dynamic system equation [10, 11].

We choose an example of q-deformed operator of conformal weight 2 through the KdV system, the associated Lax operator is given by

$$L_{q-KdV} = \partial_q^2 + u_2 \tag{61}$$

We follow the same method of the previous example, the Ansatz for operator B is as follows

$$B = \partial_q \circ L + \tilde{B} = \partial_q^3 + \eta_q(u_2)\partial_q + \partial_q(u_2) + \tilde{B} \tag{62}$$

and the associated Lax equation is given by

$$[\partial_t - B, L]_q = 0 \tag{63}$$

after a simple calculation we find

$$[L, B]_q = -\dot{u}_1 \tag{64}$$

after simplification of the Lax equation, we find the following KdV equation in its q-deformed version

$$\dot{u}_2 = \left[u_2 + \eta_q(u_2)\right]\partial_q(u_2) + \partial_q^2\left[\partial_q(u_2) + \eta_q\left(\partial_q u_2\right)\right] \tag{65}$$

In the classical limit where $q = 1$ we have $\eta_q(u_2) = u_2$ and $\partial_q(u_2) = u_2'$, however, we find the following classical KdV equation

$$\dot{u}_2 = u_2 u_2' + u_2''' \tag{66}$$

The same work can be done for a nonlinear system with equal weight 3 in the Boussinesq system and a nonlinear system with equal weight 1 in the Burgers system.

4. Moyal deformation

4.1 Fundamental notions

In this section, we treat the Moyal deformation defined on the two-dimensional phase space. Such a deformation is seen as a non-commutative formulation of the functions $f(x,p)$ where the variable x represents the coordinate of the space while p describes the coordinate of the momentum [12].

The simple vision of this deformation is given by the following main object

$$u_i(x,t) \star p^j, \tag{67}$$

which is an object of conformal weight $(i+j)$ in the noncommutative space characterized by the noncommutativity parameter θ. The star product law $\star$, defining the internal composition law of the elements in the noncommutative phase space (multiplication), which is given by the following expression

$$f(x,p) \star g(x,p) = \sum_{s=0}^{\infty} \sum_{i=0}^{s} \frac{\theta^s}{s!} (-)^i c_s^i \left(\partial_x^i \partial_p^{s-i} f\right) \left(\partial_x^{s-i} \partial_p^i g\right), \tag{68}$$

where $c_s^i = \frac{s!}{i!(s-i)!}$ is the combination parameter. We define the Moyal bracket as follows

$$\{f(x,p), g(x,p)\}_\theta = \frac{f \star g - g \star f}{2\theta}, \tag{69}$$

where θ is the parameter of the noncommutativity which is considered in our study as a constant.

4.2 Algebraic structure

Let us consider the algebra $\hat{\Sigma}(\theta)$ of all arbitrary differential momentum operators of arbitrary conformal weight m and arbitrary degrees (r,s). It is obtained as a direct sum of all operators of conformal weight m and degrees (r,s) in the following way:

$$\hat{\Sigma}(\theta) = \oplus_{r \le s} \oplus_{m \in Z} \hat{\Sigma}_m^{(r,s)}. \tag{70}$$

it is an infinite dimensional algebra which is closed under the action of the Moyal bracket [16, 17]. The subspace $\hat{\Sigma}_m^{(r,s)}$ is the algebra of differential operators of conformal weight m and degrees (r,s) with $r \leq s$, a simple prototype operator of this algebra is given by

$$\mathcal{L}_m^{(r,s)}(u) = \sum_{i=r}^{s} u_{m-i}(x) \star p^i. \tag{71}$$

the special cases of this algebra is given by: $\hat{\Sigma}_m^{(0,0)}$ which represents the ring of functions of conformal weight m and $\hat{\Sigma}_m^{(k,k)}$ which represents the algebra of momentum operators type,

$$\mathcal{L}_m^{(k,k)}(u) = u_{m-k} \star p^k. \tag{72}$$

In this algebra, we define the Residual operation (θ-Residue) by the following formula

$$\hat{Res}\left(\alpha \star p^{-1}\right) = \alpha. \tag{73}$$

and we find the following θ- Leibnitz rules

$$\begin{aligned} p^n \star f(x,p) &= \sum_{s=0}^{n} \theta^s c_n^s f^{(s)}(x,p) p^{n-s}, \\ p^{-n} \star f(x,p) &= \sum_{s=0}^{\infty} (-)^s \theta^s c_{n+s-1}^s f^{(s)}(x,p) p^{-n-s}. \end{aligned} \tag{74}$$

However, we can establish the following expressions for the Moyal bracket:

$$\begin{aligned} \{p^n, f\}_\theta &= \sum_{s=0}^{n} \theta^{s-1} c_n^s \left\{\frac{1-(-)^s}{2}\right\} f^s p^{n-s}, \\ \{p^{-n}, f\}_\theta &= \sum_{s=0}^{\infty} \theta^{s-1} c_{s+n-1}^s \left\{\frac{(-)^s - 1}{2}\right\} f^s p^{-n-s} \end{aligned} \tag{75}$$

here we present some examples of this Moyal bracket for positive values of degree (where the development is finite) and for negative values of degree (where the development is infinite)

$$\begin{aligned} \{p, f\}_\theta &= f' \\ \{p^2, f\}_\theta &= 2f'p \\ \{p^3, f\}_\theta &= 3f'^2 + \theta^2 f''' \end{aligned} \tag{76}$$

$$\begin{aligned} \{p^{-1}, f\}_\theta &= -f'^{]-2} - \theta^2 f'''^{-4} - \ldots - \theta^{2k} f^{(2k+1)} p^{-2k-2} - \ldots \\ \{p^{-2}, f\}_\theta &= -2f'^{-3} - 4\theta^2 f'''^{-5} - \ldots - (2k+2)\theta^{2k} f^{(2k+1)} p^{-2k-3} - \ldots \\ \{p^{-3}, f\}_\theta &= -3f'^{-4} - 10\theta^2 f'''^{-6} - \ldots - \frac{(2k+3)(2k+2)}{2} \theta^{2k} f^{(2k+1)} p^{-2k-4} - \ldots \end{aligned} \tag{77}$$

In the case of classical noncommutativity, the standard limit is taken such that $\theta_{limit} = 0$. Whereas in the Moyal deformation, the standard limit is shifted by (1/2) such that $\theta_l \to \theta_{limit} + \frac{1}{2}$. This shift is due to the consistent W_θ^3-Zamolodchikov algebra construction [12, 18].

4.3 Moyal deformation of Lax representation

First, we define the hierarchical structures of the Lax representation of the KdV equation in the deformed phase space, such equation named sl_n-Moyal KdV hierarchy is defined as

$$\frac{\partial \mathcal{L}}{\partial t_k} = \left\{ \left(\mathcal{L}^{\frac{k}{2}} \right)_+, \mathcal{L} \right\}_\theta. \tag{78}$$

Explicit calculations related to these hierarchies give [12, 17, 18].

$$\begin{aligned} u_{t_1} &= u', \\ u_{t_3} &= \frac{3}{2} u u' + \theta^2 u''', \\ u_{t_5} &= 5\theta^2 \left(u' u'' + \frac{1}{2} u u''' \right) + \frac{15}{8} u^2 u' + \theta^4 u^{(5)}, \\ &\dots \end{aligned} \tag{79}$$

In fact, these results are calculated for the case $n = 2$, they can be generalized for higher orders.

To realize the Lax formulation, we need to consider a noncommutative integrable system which has the Lax representation defined by the Moyal deformation of the Lax equation given by [6, 19].

$$\{L, T + \partial_t\}_\theta = 0, \tag{80}$$

The differential operator L defines the integrable system studied, consequently, it must be fixed from the start.

Note that the Lax eq. (80) is equivalent to (78) namely

$$\{L, T + \partial_t\}_\theta \equiv \left\{ L, \left(\mathcal{L}^{\frac{k}{2}} \right)_+ + \partial_{t_k} \right\}_\theta = 0, \tag{81}$$

where T is an operator analogue of $\left(\mathcal{L}^{\frac{k}{2}} \right)_+$ of conformal spin k [17].

In order to apply the θ-deformation Lax-pair generating technique, one have to make some constraints on the operator T namely:

$$T = p^n \star L^m + T', \tag{82}$$

With the previous ansatz, the problem is reduced to that of the operator T' which is determined manually so that the Lax equation is a differential equation belonging to the ring $\hat{\Sigma}^{(0,0)}$.

4.4 Moyal KdV equation

For the KdV system, The L-operator for the noncommutative equation is given by

$$L = p^2 + u(x, t), \tag{83}$$

where

$$L \in \hat{\Sigma}_2^{(0,2)} / \hat{\Sigma}_2^{(1,1)}, \tag{84}$$

for the deformed sl_2-KdV $_\theta$ system, the ansatz (82) can be written as follows

$$T = p \star L + T'. \tag{85}$$

The operator T in this case ($k = 3$), is shown to behaves as $\left(\mathcal{L}^{\frac{3}{2}}\right)_+$ with $\partial_{t_3} \equiv \frac{\partial}{\partial t_3}$. Simply algebraic computations give

$$\{L, T'\}_\theta = u'^2 - 2\theta u'' p + \frac{\dot{u}}{2\theta} + \left(uu'^2 u'''\right), \tag{86}$$

for T', we have to consider the following Ansatz [13]:

$$T' = X \star p + Y, \tag{87}$$

where X and Y are arbitrary functions on u and its derivatives. With $T' = Xp - \theta X' + Y$ and by performing straightforward computations we have

$$\begin{aligned} \{L, T'\}_\theta = (\{u, X\}_\theta - 2\theta X'' + 2Y')p + \\ 2X'^2 + (-Xu' - \theta\{u, X'\} + \{u, Y\}) \end{aligned} \tag{88}$$

Identifying (86) and (88), leads to the following constrains on the functions X and Y

$$X = \frac{1}{2}u_2 + a, \quad Y = -\frac{1}{2}\theta u_2' + b, \tag{89}$$

with the following nonlinear differential equation

$$-\frac{\dot{u}}{2\theta} = \frac{3}{2}uu'^2 u'''. \tag{90}$$

where the constants a and b are to be omitted for a matter of simplicity. The last equation is noting but the sl_2 KdV$_\theta$ equation. This deformed equation contains also a non linear term $\frac{3}{2}uu'$. We have to underline that the sl_2 KdV$_\theta$ equation obtained through this Lax method belongs to the same class of the KdV equation derived in [12, 16] namely

$$\dot{u} = \frac{3}{2}uu'^2 u'''. \tag{91}$$

In fact, performing the following scaling transformation $\partial_{t_3} \to -2\theta\partial_{t_3}$ we recover exactly the KdV equations.

5. Conclusion

The Korteweg-de Vries equation was formulated to explain the solitary water waves observed by J. Scott Russell in the Edinburgh Glasgow Canal. It represents a nonlinear equation in a (1 + 1) space–time. As we show, It has soliton solutions. But at the time of its formulation, nothing was known about the integrability of the equation.

We set up the KdV equation as a Hamiltonian system and we analyzed the integrability of this system. In order to generalize the KdV equation, we have discussed the solutions of the KdV equation in the q-deformed case through the Lax pair method.

The θ-deformation of the KdV system shows, among others, the consistency of the Moyal momentum formulation in describing integrable systems and the associated Lax pair generating technics in the same way as the successful GD formulation [17–19].

Classification

PACS number(s): 35Q53; 37 K10; 81R60; 81 T75; 05A30

Author details

Abderrahman El Boukili[1*], Hicham Lekbich[1], Tahir Toghrai[1], Najim Mansour[1] and Moulay Brahim Sedra[2]

1 Faculty of Sciences and Techniques, Team of Theoretical Physics and Modelisation, Moulay Ismail University, Errachidia, Morocco

2 Faculté des Sciences, Département de Physique, LPMS, Université Ibn Tofail, Laboratoire de physique des Matériaux et Subatomique, Kenitra, Morocco

*Address all correspondence to: a.elboukili@umi.ac.ma

References

[1] Albert TD, Ruhlig K, Pakuliak S, von Gehlen G. Integrable Structures of Exactly Solvable Two-Dimensional Models of Quantum Field Theory. Netherlands: Springer; 2001

[2] Das A. Integrable Models. World Scientific Publishing Company; 1989

[3] Bobenko AI, Reyman AG, Semenov-Tian-Shansky MA. The Kowalewski top 99 years later: A Lax pair, generalizations and explicit solutions. Communications in Mathematical Physics. 1989;**122**(2): 321-354

[4] Lax P. Integrals of nonlinear equations of evolution and solitary waves. Communications on Pure and Applied Mathematics. 1968;**21**(5): 467-490

[5] Fomenko AT. Symplectic geometry, 2e édition. CRC Press; 1995. ISBN-10: 2881249019

[6] Sergyeyev A. New integrable (3+1)-dimensional systems and contact geometry. Letters in Mathematical Physics. 2018;**108**(2):359-376. DOI: 10.1007/s11005-017-1013-4

[7] Miura RM, Equation K-d V, Generalizations I. A remarkable explicit nonlinear transformation. Journal of Mathematical Physics. 1968;**9**:1202-1204. DOI: 10.1063/1.1664700

[8] Sardanashvily G. Handbook of Integrable Hamiltonian Systems. URSS; 2015

[9] Biedenharn LC, Lohe MA. Quantum Group Symmetry and Q-Tensor Algebras. Singapore: World Scientific; 1995

[10] El Boukili A, Sedra MB. On the algebra of Q-deformed Pseudo differential operators. ISRN High Energy Physics. 2012;**2012**:503621. DOI: 10.5402/2012/503621

[11] Frenkel E. Deformations of the KdV hierarchy and related soliton equations. IMRN. 1996;**1996**:55-76

[12] Boulahoual A, Sedra MB. The Moyal momentum algebra applied to (theta)-deformed 2d conformal models and KdV-hierarchies. Chinese Journal of Physics. 2005;**43**:408

[13] Hamanaka M, Toda K. Towards noncommutative Integrable equations. In: Proceedings of Institute of Mathematics of NAS of Ukraine. Vol. 50. EconfC0306234; 2004. pp. 404-411

[14] Mussardo G. Statistical Field Theory: An Introduction to Exactly Solved Models of Statistical Physics. Oxford University Press; 2010

[15] Andrews GE. Regional conference series in mathematics 66. In: Q-Series: Their Development and Applications in Analysis, Number Theory, Combinatorics, Physics and Computer Algebra. Providence, RI: American Mathematical Society; 1986

[16] Hamanaka M. Noncommutative Solitons and Quasideterminants. Physica Scripta. 2014;**89**:038006

[17] Zamolodchikov AB. Tetrahedron equations and the relativistic S-matrix of straight-strings in 2 + 1-dimensions, Integrable Systems in Statistical Mechanics. In: D'Ariano G, Montorsi A, Rasetti M, editors. Series on Advances in Statistical Mechanics. Vol. 1. World Scientific; 1985 178 pp

[18] Sedra MB. On the huge lie superalgebra of pseudo superdifferential

operators and super KP hierarchies. Journal of Mathematical Physics. 1996; **37**:3483. DOI: 10.1063/1.531587

[19] Saidi EH, Sedra MB. On the Gelfand dickey algebra GD(SLn) and the Wn-symmetry. I. the bosonic case. Journal of Mathematical Physics. 1994;**35**:3190. DOI: 10.1063/1.530461

Chapter 3

Online Canonical Polyadic Decomposition: Application of Fluorescence Tensors with Nonnegative Orthogonality and Sparse Constraint

Isaac Wilfried Sanou, Xavier Luciani, Roland Redon and Stéphane Mounier

Abstract

The canonical polyadic decomposition (CPD) is now widely used in signal processing to decompose multi-way arrays. In many applications, it is important to add constraints to quickly converge on an optimal solution. In contrast to classical CPD, we then focus on online CPD. In this context, the number of relevant factors is usually unknown and can vary with time. We propose two algorithms to compute the online CPD based on sparse dictionary learning. We also introduce an application example in environmental sciences and evaluate the performances of the proposed approaches in this context on real data.

Keywords: third order tensor decomposition, online tensor decomposition, PARAFAC, rank variation, fluorescence spectroscopy

1. Introduction

In many fields such as psychometric [1], data mining [2], neuroscience [3], chemometric [4], telecommunications [5], computer vision [6], and biomedical image processing [7], use canonical polyadic decomposition (CPD) also known as PARAllel FActor analysis (PARAFAC).

The data from these fields can be put in a multidimensional data array (or tensor). Then, the CPD permits to decompose this array into factors (which is composed of the number of components or rank depending to the application) that can be interpreted by the user.

CPD algorithms can be summarized in three mains methods:

- *Direct methods.* These methods are based on algebraic computation such as Direct TriLinear Decomposition (DTLD) [8], SEmi-algebraic framework for approximate

CP decompositions via SImultaneous matrix diagonalization (SECSI) [9], and DIrect AlGorithm for canonical polyadic decomposition (DIAG) [10].

- *Alternating methods.* These methods are based on least square methods, and we can cite alternating least square (ALS) [1], hierarchical alternating least square (HALS) [11–13], and alternating direction method of multipliers (ADMM) [14].

- *Descent methods.* These methods are based on traditional gradient descent. We can cite some algorithms using stochastic gradient descent [15–17] or Nadam optimizer [18].

By considering these algorithms, we can say that the rank of the decomposition, namely the relevant number of factors, is known, but we find in practice that this is not the case. Indeed, this value is unknown so we make an estimate on the order of magnitude while a bad choice can have an impact on the factorial estimates [19]. Traditionally, two opposing approaches are used to address this situation.

- *Rank estimation.* Rank estimation consists in estimating the appropriate CPD rank before the decomposition. Among these methods, we can cite the CORe CONsistency DIAgnostic (CORCONDIA) [20], split half validation [21], or AutoTen [22]. Though these approaches do not always allow to clearly decide between several possible rank values.

- *Overfactoring.* In this method, an overestimated value is chosen, i.e., one higher than the current rank and the CPD algorithms are designed to produce additional factors with zero contribution [19]. Thus, in contrast, overfactoring is a posterior rank estimation method because the appropriate rank value is inferred from the CPD.

At this stage, we have talked about offline CPD. However, in this chapter, we are interested in online CPD. In online CPD, the data increases with the time, and the decomposition must follow this augmentation [23].

In the context of fluorescence data and environmental sciences, the online CPD can be summarized as follow:

- 3-way array is used. This tensor is built by concatenation of a new matrix on the last mode. At each time we got a tensor call sub-tensor as shown in **Figure 1**.

- In practice, some components or the rank of the decomposition can change at interval time from one sub-tensor to another. This phenomenon can be seen as appear and/or disappear of component. We talk about the rank variation but these variations are unknown.

- The goal of online CPD in this context is to update at each new time interval the CPD factors using the factors estimated previously without performing the CPD of the whole.

In the literature, some online CPD algorithms exist. For example, in [23], the authors introduced two adaptive algorithms: simultaneous diagonalization tracking (SDT) which tracks the SVD of the unfolded tensor and recursive least squares tracking (RLST). In [24], grid-based tensor factorization algorithm (GridTF) is introduced for the large tensor. Also, in [25] authors proposed OnlineCP based on ALS

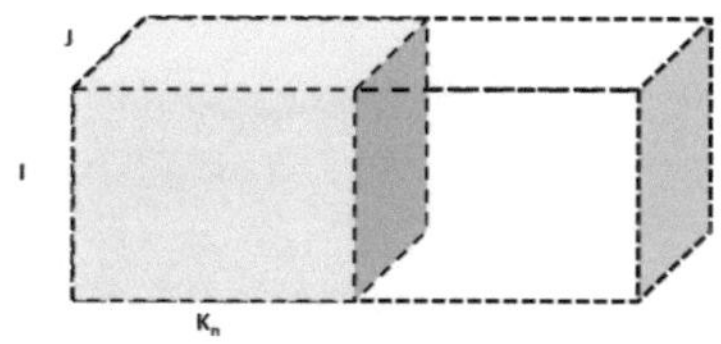

Figure 1.
Example of a sub-tensor.

algorithms. In [26] authors introduced an online Tucker decomposition with a fixed rank. This list is not exhaustive, and more online algorithms can be found in [27].

For the rest of the chapter, we propose an approach based on sparse dictionary learning to compute the CPD with sparsity, nonnegativity, and orthogonality constraint. These constraints permit to handling unknown rank variation over time and the better convergence. Considering the most general case, we make no assumptions about these variations. The goal of the proposed approach is to combine dictionary learning with LASSO [28] in a simple but appropriate way for nonnegative online CPD. We have selected the factors of the two fixed modes from two dictionaries which are learned and updated throughout the online process. The appropriate number of vectors is selected from a sparsity constraint on the two atoms matrices (i.e., the CPD rank). We derived from this solution three different algorithms (one for offline CPD and the rest for online CPD). They differ in the way that the dictionaries are updated from one sub-tensor to another. Next, we propose an experiment semi-controlled to obtain real fluorescence data online with rank variations. We provide out an evaluation od our approaches with the state of the art.

Notations are introduced while recalling some basic definitions and the problem of online CPD in the particular context of fluorescence spectroscopy and environmental sciences in Section 2. In Section 3, the new proposed approach is described in the form of three algorithms for computing the nonnegative CPD of a third-order tensor. Two of these algorithms were already presented in a previous conference paper [18] and this journal paper [27]. In Section 4, an experimental design is proposed to evaluate the approaches. Results and comparisons with reference approach are provided and discussed. Section 5 concludes.

2. Problem formulation

We denote tensors and sub-tensors with letter $\mathcal{T}$ and are of size $(I \times J \times K)$ meaning that $\mathcal{T}$ gather K matrix of size $(I \times J)$ on its last mode.

2.1 Offline CPD: Example of a fluorescence tensor

In environmental sciences, especially for fluorescence data tensors the third-order tensor is most used. The CPD of this tensor can be used in order to characterize fluorescent dissolved organic matters (FDOM) in natural water samples [21, 27, 29–31] (see **Figure 2**).

Fluorescence data tensors are a set of 2D signals called emission and excitation matrices (EEMs) of fluorescence measured from a set of samples. Each sample is then a mixture of an unknown number of fluorescent components (fluorophores), and each part of an EEM corresponds to the fluorescence intensity of one sample at a

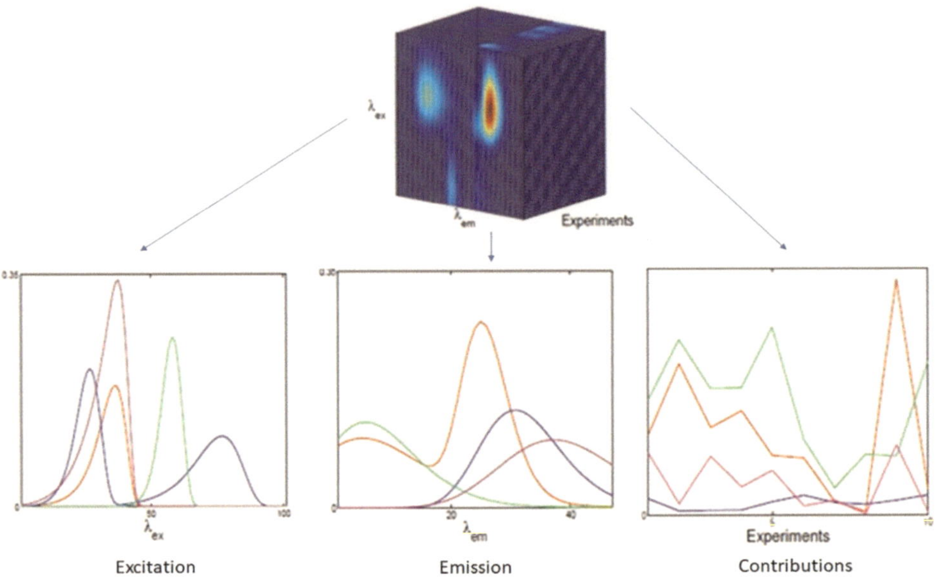

Figure 2.
Example of the rank 4 CPD of the fluorescence tensor [18, 27].

given couple of excitation and emission wavelengths. At low concentrations, the nonlinear model based on the Beer–Lambert law can be linearized, and it then follows the CPD model [32]. To recover each individual emission and excitation spectra of the fluorophores present in the different samples along with their respective contributions can be found using the CPD. In some applications like in our case, the decomposition factors have physical meaning and are known to be nonnegative. This constraint allows better convergence especially when the columns of the factors are collinear. Nonnegativity constraint can be imposed on the factors during optimization by projecting the values of each factor matrix on $\mathbb{R}^+$ [33]. This method of applying nonnegativity constraint sometimes shows its limits in certain cases due to the change of scale. In [34], the authors propose to solve this problem by involving two factors, ϵ and $\theta \in \{-1; 1\}$, which multiply each factors. Nonnegativity can also be imposed by choosing an optimization method [14, 17, 35] or even using an exponential variable change [31]. In the literature, the projection method is mostly used such as in approach [14, 17, 18, 33, 36, 37]. In signal or data processing, the nonnegative CPD is usually used as a mathematical model to fit a data tensor $\mathcal{T}$ of size (I, J, K):

$$\forall i, j, k, \quad \mathcal{T}_{i,j,k} \simeq \widehat{\mathcal{T}}_{i,j,k}(\mathbf{A}, \mathbf{B}, \mathbf{C}) = \sum_{r=1}^{R} \mathbf{A}_{ir}\mathbf{B}_{jr}\mathbf{C}_{kr} \tag{1}$$

The matrices $\mathbf{A} \in \mathbb{R}_+^{I \times R}$, $\mathbf{B} \in \mathbb{R}_+^{J \times R}$ and $\mathbf{C} \in \mathbb{R}_+^{K \times R}$ are named the factor matrices. R is the CPD rank. Each column of the factor matrices **A**, **B**, and **C**, defines the CPD factors. We can distinguish two values of R if the CPD factors have a physical meaning:

- The define value of the tensor $\mathcal{T}$.

- The maximal value of R for which all the factors have a physical meaning. We name this value the physical rank of $\mathcal{T}$. It is usually much smaller than the tensor rank.

In the considered application, these matrices are usually full column rank so that the Kruskal condition [38] is fulfilled and guarantees that the decomposition is unique up to trivial scaling and permutation indeterminacy.

The *euclidia* distance F is the most used to the reconstruction error. With nonnegative constraint, the offline CPD problem is thus given by

$$\begin{aligned} &\min\left\{F(\mathbf{A},\mathbf{B},\mathbf{C}) = \|\mathcal{T} - \widehat{\mathcal{T}}(\mathbf{A},\mathbf{B},\mathbf{C})\|_F^2\right\} \\ &s.t.\ \ \mathbf{A}\geq 0, \mathbf{B}\geq 0, \mathbf{C}\geq 0 \end{aligned} \tag{2}$$

2.2 Fluorescence tensor in online CPD

In the online case, tensor $\mathcal{T}$ can be seen as

- A composition of **sub-tensors** $\mathcal{T}_n$ of size (I, J, K_n) acquired at various time intervals.
- Each sub-tensor can have zero matrix in common between last and the present sub-tensor. We call this case partition with *nooverlapping.*
- In opposite case, we talk about overlapping case. As explained on **Figure 3**, two consecutive sub-tensors share some matrices, *i.e*, in this case, $\mathcal{T}_n$ also contains the last η EEMs of $\mathcal{T}_{n-1}$. Sub-tensor overlapping should help the factor estimations but then the online decomposition becomes slower because we find ourselves with a larger number of sub-tensors.
- Rank variation or fluorophore appearance and disappearance can be explained by sea currents or pollution events, natural degradation ... , in a natural marine environment.

The physical rank of sub-tensor $\mathcal{T}_n$ will be denoted $\tilde{R}_n$ in this chapter, and we recall that it is unknown. The online CPD problem with rank variation has been clearly described in [18, 27, 39].

In short, the online CPD consists by solving Eq. (2) for each sub-tensor $\mathcal{T}_n$ with the nonnegative constraint. We speak of: NonNegative Online Canonical Polyadic Decomposition (NNOnline CPD). The factor matrices estimated at time t_{n-1} with the sub-tensor $\mathcal{T}_{n-1}$ are used to update the sub-tensor $\mathcal{T}_n$ in order to reduce the computational cost.

3. Algorithms for computing the nonnegative online canonical polyadic decomposition with sparsity and orthogonality constraint

3.1 NNCPD for the first tensor acquired at time t_0

The first sub-tensor acquired at time t_0 is denoted $\mathcal{T}_0$. At this time, we do not know the value of $\tilde{R}_0$ but we suppose that, $\forall n, \tilde{R}_n < < R$. We aim to rewrite the matrices **A** and **B** as the product of sparse dictionaries $\mathbf{D}^A$ and $\mathbf{D}^B$ by atoms $\mathbf{V}^A$ and $\mathbf{V}^B$:

$$\mathbf{A} = \mathbf{D}^A\mathbf{V}^A \text{ and } \mathbf{B} = \mathbf{D}^B\mathbf{V}^B \tag{3}$$

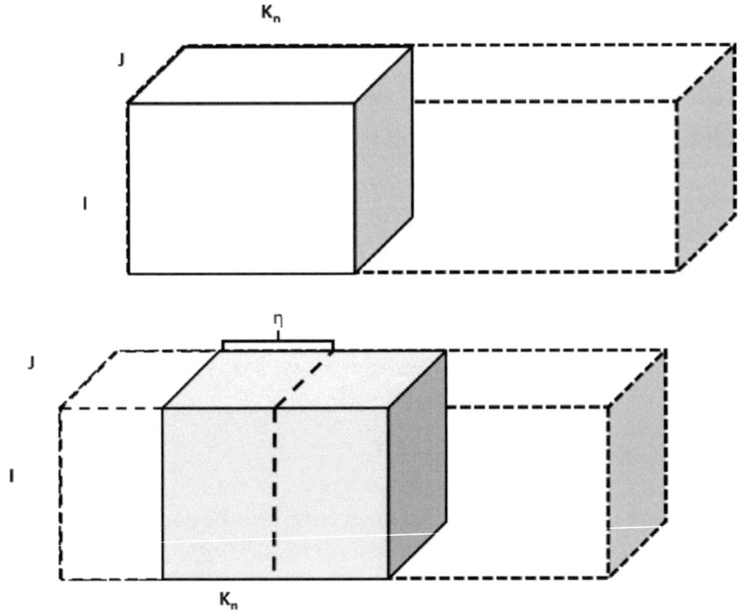

Figure 3.
Example of a sub-tensor.

with $\mathbf{D}^A(I, R)$ and $\mathbf{D}^B(J, R)$, respectively, and $\mathbf{V}^A(R, R)$ and $\mathbf{V}^B(R, R)$. We expect that $\mathbf{D}^A$ and $\mathbf{D}^B$ contain $\tilde{R}$ true components on their columns. The factors without physical meaning are $R - \tilde{R}$. While $\mathbf{V}^A$ and $\mathbf{V}^B$ have $R - \tilde{R}$ null columns and that the other columns form generalized permutation matrix. For instance, for $R = 3$ and $\tilde{R}_0 = 2$, we could have ideally

$$\mathbf{A} = \overbrace{\begin{pmatrix} 9 & 7 & 2 \\ 1 & 4 & 2 \\ 2 & 1 & 2 \end{pmatrix}}^{\mathbf{D}^A} \cdot \overbrace{\begin{pmatrix} 0.8 & 0 & 0 \\ 0 & 0 & 0 \\ 0 & 0.9 & 0 \end{pmatrix}}^{\mathbf{V}^A} = \begin{pmatrix} 7.2 & 1.8 & 0 \\ 0.8 & 1.8 & 0 \\ 1.6 & 1.8 & 0 \end{pmatrix}$$

We can easily estimate $\tilde{R_0}$ by counting the number of non-null columns of $\mathbf{A}$ or $\mathbf{B}$. A similar problem has been studied in depth by Cohen and Gillis in [40] but for the case in which the dictionary is known and the atom is a selection matrix. To have zeros columns on the atoms, we thus promote the sparsity constraints on the matrices $\mathbf{V}^A$ and $\mathbf{V}^B$. In this purpose, we will aim at minimizing their $L_{1,1}$ norm. Regarding the sparsity constraint, it is generally used for sparse tensors or to ensure the overestimation of the rank, as mentioned above. This constraint is imposed according to the goal we are looking for, and in our case, it is the overestimation of the rank. The constraint can be applied to the matrices factors in the form of a regularization term:

$$\min_x \overbrace{F(x)}^{\text{fonction}} + \underbrace{R(x)}_{\text{regularization}} \tag{4}$$

$F(x)$ corresponds here to Eq. (2) and is the attachment part. $R(x)$ is the regularization part on the matrix **a, b,** and **c**.

$L_{1,1}$ norm is a classically used to promote the sparsity [41]. Indeed, it is more tractable than the $L_{0,0}$ norm. It has been used in similar contexts in [17, 35, 36, 40]. We can thus rewrite our problem as a LASSO regression problem by adding two penalty terms to the cost function (see Appendix A).

For $\mathcal{T} = \mathcal{T}_0$, we solve

$$\begin{aligned} &\min\{F_1(\mathbf{D}^A, \mathbf{V}^A, \mathbf{D}^B, \mathbf{V}^B, \mathbf{C})\} \quad s.t. \quad \mathbf{D}^A, \mathbf{V}^A, \mathbf{D}^B, \mathbf{V}^B, \mathbf{C} \geq 0 \\ &\text{where } F_1 = \frac{1}{2}\|\mathcal{T} - \widehat{\mathcal{T}}(\mathbf{D}^A\mathbf{V}^A, \mathbf{D}^B\mathbf{V}^B, \mathbf{C})\|_F^2 + \alpha\|\mathbf{V}^A\|_{1,1} + \alpha\|\mathbf{V}^B\|_{1,1} \end{aligned} \tag{5}$$

$\alpha > 0$ is a penalty term.

This approach is named sparse nonnegative CPD (SNNCPD). We resort to a stochastic gradient descent (SGD) algorithm called Nadam [42, 43] in order to solve Eq. (5). Details of the Nadam algorithm are given in Appendix B. In order to ensure the nonnegativity of matrix entries, all the element of matrices $\mathbf{D}^A, \mathbf{D}^B, \mathbf{C}, \mathbf{V}^A$, and $\mathbf{V}^B$ are projected on $\mathbb{R}_+$ at each iteration [44]. We can thus differentiate the $L_{1,1}$ norm, and the gradients with respect to the different variables are given by

$$\begin{aligned} \frac{\partial F_1}{\partial \mathbf{V}^A} &= -(\mathbf{D}^A)^\top \left(\mathbf{T}_1 - \mathbf{D}^A\mathbf{V}^A(\mathbf{L}_1)^\top\right)\mathbf{L}_1 + \alpha \mathbf{1}_{R,R} \\ \frac{\partial F_1}{\partial \mathbf{V}^B} &= -(\mathbf{D}^B)^\top \left(\mathbf{T}_2 - \mathbf{D}^B\mathbf{V}^B(\mathbf{L}_2)^\top\right)(\mathbf{L}_2) + \alpha \mathbf{1}_{R,R} \\ \frac{\partial F_1}{\partial \mathbf{D}^A} &= -\left(\mathbf{T}_1 - \mathbf{D}^A\mathbf{V}^A(\mathbf{L}_1)^\top\right)\mathbf{L}_1(\mathbf{V}^A)^\top \\ \frac{\partial F_1}{\partial \mathbf{D}^B} &= -\left(\mathbf{T}_2 - \mathbf{D}^B\mathbf{V}^B(\mathbf{L}_2)^\top\right)(\mathbf{L}_2)(\mathbf{V}^B)^\top \\ \frac{\partial F_1}{\partial \mathbf{C}} &= -\left(\mathbf{T}_3 - \mathbf{C}(\mathbf{L}_3)^\top\right)(\mathbf{L}_3) \end{aligned} \tag{6}$$

Matrices $\mathbf{T}_1, \mathbf{T}_2, \mathbf{T}_3$ are obtained by unfolding the tensor $\mathcal{T}$ with respect to the first, second, and third modes, respectively. $\mathbf{L}_1 = \mathbf{C} \odot (\mathbf{D}^B\mathbf{V}^B), \mathbf{L}_2 = \mathbf{C} \odot (\mathbf{D}^A\mathbf{V}^A)$ and $\mathbf{L}_3 = (\mathbf{D}^B\mathbf{V}^B) \odot (\mathbf{D}^A\mathbf{V}^A)$. The matrices $\mathbf{V}^A$ and $\mathbf{V}^B$ are initialized as the identity matrix. The matrices $\mathbf{D}^A$ and $\mathbf{D}^B$ and $\mathbf{C}$ are initialized with nonnegative random values. The different steps of SNNCPD are summarized in Algorithm 1.

Algorithm 1. Sparse NonNegative Canonical Polyadic Decomposition (SNNCPD). Initialization step

---

- **Input:** $\mathcal{T}$, R overestimated

while *a convergence criterion is not reached* **do.**
Update the dictionaries $\mathbf{D}^A, \mathbf{D}^B$, *and* $\mathbf{C}$ *using Nadam optimizer.*
Update the atoms $\mathbf{V}^A$ *and* $\mathbf{V}^B$ *using Nadam optimizer.*
end while

- **Output:** $(\mathbf{D}_a, \mathbf{D}_b, \mathbf{C}, \mathbf{V}_a, \mathbf{V}_b)$

3.2 Online step: NN-CPD for the first tensor acquired at time t_n with $n > 0$

In this step, we assume that the factors matrices $\mathbf{A}_n$ and $\mathbf{B}_n$ are well estimated.

3.2.1 Algorithm 1

In this approach, we allow some modification of the factors estimated at time t_{n-1} thanks to linear combinations as explained below. We first replace all the null columns of $\mathbf{A}_{n-1}$ and $\mathbf{B}_{n-1}$ by columns of random numbers drawn from the standard normal distribution. We then look for $\mathbf{A}_n$ and $\mathbf{B}_n$ as

$$\mathbf{A}_n = \mathbf{U}^A \mathbf{A}_{n-1} \mathbf{V}^A \text{ and } \mathbf{B}_n = \mathbf{U}^B \mathbf{B}_{n-1} \mathbf{V}^B \tag{7}$$

The matrices $\mathbf{U}^A$ and $\mathbf{U}^B$ are square with sizes I and J, respectively. Here, $\mathbf{U}^A \mathbf{A}_{n-1}$ and $\mathbf{U}^B \mathbf{B}_{n-1}$ are our dictionaries, while $\mathbf{V}^A$ and $\mathbf{V}^B$ matrices are still sparse atoms. Therefore, the optimization problem becomes

$$\begin{aligned} &\min\{F_2(\mathbf{U}^A, \mathbf{V}^A, \mathbf{U}^B, \mathbf{V}^B, \mathbf{C})\} \;\; s.t \;\; \mathbf{U}^A, \mathbf{U}^B, \mathbf{C}, \mathbf{V}^B, \mathbf{V}^B \geq 0 \\ &\text{with } F_2 = \frac{1}{2} \|\mathcal{T} - \widehat{\mathcal{T}}(\mathbf{U}^A \mathbf{A}_{n-1} \mathbf{V}^B, \mathbf{U}^B \mathbf{B}_{n-1} \mathbf{V}^B, \mathbf{C})\|_F^2 \\ &+\alpha \|\mathbf{V}^A\|_1 + \alpha \|\mathbf{V}^B\|_1 \end{aligned} \tag{8}$$

We call this algorithm online sparse and nonnegative CPD (OSNCPD). The different steps are.

summarized in Algorithm 2. Most details can be found in [18, 27].

Algorithm 2. Online Sparse and NonNegative CPD

- **STEP 1: Initialization phase**

Input: $\mathcal{T}_0$, *R overestimated.*
Solve Eq. (5) *with* $\mathcal{T} = \mathcal{T}_0$.
Compute matrices $\mathbf{A}_0$ *and* $\mathbf{B}_0$ *using Eq.* (3).
Output: $\mathbf{A}_0, \mathbf{B}_0$ *and* $\mathbf{C}_0$

- **STEP 2: Online phase at time t_n with $n > 0$**

Input: $\mathcal{T}_n$, *and* $\mathbf{A}_{n-1}, \mathbf{B}_{n-1}$.
Fill null columns of $\mathbf{A}_{n-1}, \mathbf{B}_{n-1}$ *by random numbers and Solve* Eq. (8).
Update matrices $\mathbf{A}_n$ *and* $\mathbf{B}_n$ *using Eq.* (7).
Output: $(\mathbf{A}_n, \mathbf{B}_n, \mathbf{C}_n$ *and* $\tilde{R}_n)$

- **STEP 3: Return to Step 2 with n = n + 1**

The advantage of this approach is therefore its flexibility. The inconvenient is that the matrices to optimize are larger which make the algorithm more expensive in time.

We call this algorithm online sparse and nonnegative CPD (OSNCPD). The different stages are summarized in the Algorithm 2.

3.2.2 *Algorithm 2*

By taking the Eq. (7) as well as the cost function 8, the optimization of the algorithm becomes costly because of the dictionaries $\mathbf{U}^A$ and $\mathbf{U}^B$. This is caused by the fact that no constraint is applied to these matrices to restrict the research space. In this approach, we will apply an orthogonality constraint on these dictionaries. The orthogonality constraint on the factors is also useful in certain applications [26, 45]. It is possible to apply the orthogonality constraint to the factors by using the Eq. 4. If, for example, $\mathbf{A}$ must be orthogonal then: $\mathbf{I}_A = \mathbf{A}^t\mathbf{A}$. We therefore seek to minimize

$$\min_{\mathbf{A},\mathbf{B},\mathbf{C}} \left\| \mathcal{T} - \widehat{\mathcal{T}}(\mathbf{A},\mathbf{B},\mathbf{C}) \right\|_F^2 + \left\|\mathbf{A}^T\mathbf{A} - \mathbf{I}_A\right\|_F + \left\|\mathbf{B}^T\mathbf{B} - \mathbf{I}_B\right\|_F + \left\|\mathbf{C}^T\mathbf{C} - \mathbf{I}_C\right\|_F \tag{9}$$

To impose orthogonality constraint on the factors matrices for $\mathcal{T} = \mathcal{T}_N$, the cost function becomes

$$\begin{aligned}
&\min\{E_2(\mathbf{U}_n^A, \mathbf{V}_n^A, \mathbf{U}_n^B, \mathbf{V}_n^B, \mathbf{C})\} \\
&\text{où } E_2 = \frac{1}{2}\|\mathcal{T} - \widehat{\mathcal{T}}(\mathbf{U}_n^A\mathbf{A}_{n-1}\mathbf{V}_n^B, \mathbf{U}_n^B\mathbf{B}_{n-1}\mathbf{V}^B, \mathbf{C})\|_F^2 + \alpha\|\mathbf{V}_n^A\|_{1,1} + \alpha\|\mathbf{V}_n^B\|_{1,1} + \\
&\|\mathbf{I}_A - (\mathbf{U}_n^A)^T\mathbf{U}_n^A\|_F^2 + \|\mathbf{I}_B - (\mathbf{U}_n^B)^T\mathbf{U}_n^B\|_F^2
\end{aligned} \tag{10}$$

The gradient of the cost Eq. (10) is given by

$$\begin{aligned}
\frac{\partial E_2}{\partial \mathbf{V}_n^A} &= -(\mathbf{U}_n^A\mathbf{A}_{n-1})^\top\left(\mathbf{T}^A - \mathbf{U}_n^A\mathbf{A}_{n-1}\mathbf{V}_n^A(\mathbf{Z}^A)^\top\right)\mathbf{Z}^A + \alpha\mathbf{1}_R \\
\frac{\partial E_2}{\partial \mathbf{V}_n^B} &= -(\mathbf{U}_n^B\mathbf{B}_{n-1})^\top\left(\mathbf{T}^B - \mathbf{U}_n^B\mathbf{B}_{n-1}\mathbf{V}_n^B(\mathbf{Z}^B)^\top\right)\mathbf{Z}^B + \alpha\mathbf{1}_R \\
\frac{\partial E_2}{\partial \mathbf{U}_n^A} &= -\left(\mathbf{T}^A - \mathbf{U}_n^A\mathbf{A}_{n-1}\mathbf{V}_n^B(\mathbf{Z}^A)^\top\right)\mathbf{Z}^A(\mathbf{V}_n^A)^\top(\mathbf{A}_{n-1})^\top - \left(2\mathbf{U}_n^A\left(\mathbf{I}_A^\top - (\mathbf{U}_n^A)^\top\mathbf{U}_n^A\right)\right) + \\
&\left(2\mathbf{U}_n^A\left(\mathbf{I}_A\mathbf{U}_n^A - (\mathbf{U}_n^A)^\top\mathbf{U}_n^A\right)\right) \\
\frac{\partial E_2}{\partial \mathbf{U}_n^B} &= -\left(\mathbf{T}^B - \mathbf{U}_n^B\mathbf{B}_{n-1}\mathbf{V}^B(\mathbf{Z}^B)^\top\right)\mathbf{Z}^B(\mathbf{V}_n^B)^\top(\mathbf{B}_{n-1})^\top - \left(2\mathbf{U}_n^B\left(\mathbf{I}_B^\top - (\mathbf{U}_n^B)^\top\mathbf{U}_n^B\right)\right) + \\
&\left(2\mathbf{U}_n^B\left(\mathbf{I}_B\mathbf{U}_n^B - (\mathbf{U}_n^B)^\top\mathbf{U}_n^B\right)\right) \\
\frac{\partial E_2}{\partial \mathbf{C}} &= -\left(\mathbf{T}^C - \mathbf{C}(\mathbf{Z}^C)^\top\right)\mathbf{Z}^C
\end{aligned} \tag{11}$$

The advantage of this approach is its speed of convergence because we reduce the research space for our dictionaries. For example, if $\mathbf{A}_n$ is similar to $\mathbf{A}_{n-1}$, the matrix $\mathbf{U}_n^A$ is a diagonal matrix. We call this algorithm orthogonality online sparse CPD (OOSCPD). The different steps are similar with the Algorithm 2.

4. Experiment for fluorescence real data acquisition

We plan an experiment to obtain real data acquisition. Four well-known fluorophores are injected on a reservoir at different time intervals under quasi-real conditions. At all, we get 50 matrices or EEMs os size [21, 36] online with a spectro-fluorimeter *Hitachi F7000*. The corresponding fluorescence tensor was partitioned into four successive sub-tensors ($\mathcal{T}_0 \dots \mathcal{T}_3$). Two kinds of partitions are considered. In the first partition (50% overlapping), each sub-tensor contains 20 EEMs, and two consecutive sub-tensors have ten EEMs in common. In the second partition (no overlapping, see **Figure 3**), sub-tensors have no EEMs in common. The initialization tensor $\mathcal{T}_0$ contains 20 EEMs. The other sub-tensors contain ten EEMs. During the acquisition, an extra fluorophore appears in all the EEMs due to the experimental device. A preliminary study showed that this extra fluorophore can be represented by a single additional factor in the CPD model. Thus, the rank of the sub-tensors can be 3, 4, or 5. We recall that SNNCPD is used for sub-tensors $\mathcal{T}_0$ in initialization phase.

4.1 Results and discussions

OSNCPD has already been compared with others online algorithms in [18, 27]. We compare the values of the physical ranks estimated by the two online approaches (OSNCPD, OOSCPD) with NN-CPD algorithm proposed in [31] and the actual values. Results are reported in **Tables 1** and **2** for the 50% overlapping case and no overlapping case, respectively. The penalty coefficient term (α) used in our two algorithms and in NN-CPD is also given. We can see that NN-CPD overestimated the rank in the last sub-tensors in both cases and this means that NN-CPD creates

Symbols	Definition
$\mathcal{T}, \mathcal{T}_n$	Tensor, sub-tensor at time t_n
A, $\mathbf{A}^{\mathrm{T}}$, a, a	Matrix, transposed matrix, column vector, scalar
$\mathbf{1}_R$	A matrix (R, R) of 1
$\mathbf{I}_A$	Identity matrix with size of matrix A
$\mathbb{R}$	Set of real numbers
$\Vert\cdot\Vert_F, \Vert\cdot\Vert_{1,1}$	Frobenius norm, $L_{1,1}$ norm
$\mathbf{A} \geq 0$	Means that all the elements of matrix A are nonnegative
$\tilde{R}_n$, R_n	Physical rank of the sub-tensor $\mathcal{T}_n$, CPD rank of the sub-tensor
$\odot$	Khatri-Rao product

Table 1.
Main notations used in the paper.

Sub-tensor	0	1	2	3	α
True rank	3	4	5	5	—
NN-CPD	3	4	7	10	0.5
OSNCPD & OOSCPD	3	4	5	5	1

Table 2.
Rank estimation in the overlapping case.

duplicate factors. In contrast, our approaches (OSNCPD and OOSCPD) find the correct rank.

When the null columns of $\hat{\mathbf{A}}$ and $\hat{\mathbf{B}}$ and the corresponding columns of $\hat{\mathbf{C}}$ are removed, we can compare the factor matrices estimated by our algorithms with the true factors by means of normalized root mean squared errors:

$$E_A = \frac{\|\mathbf{A} - \hat{\mathbf{A}}\|_F}{\|\mathbf{A}\|_F}, \quad E_B = \frac{\|\mathbf{B} - \hat{\mathbf{B}}\|_F}{\|\mathbf{B}\|_F} \text{ and } E_C = \frac{\|\mathbf{C} - \hat{\mathbf{C}}\|_F}{\|\mathbf{C}\|_F} \tag{12}$$

Results are reported in **Tables 3** and **4** for the 50% overlapping case and no overlapping case, respectively **Table 5**.

In addition, to give some physical meaning to these error terms that may seem high, we have plotted in **Figure 4** the factors obtained from our approach in the sub-tensor $\mathcal{T}_4$ along with the true factors. We observe a good agreement between the true and estimated factors.

In the overlapping case, both algorithms give similar results for sub-tensors, and this can be explained that the algorithms use past information. Also, these are close to those obtained from the initialization sub-tensor $\mathcal{T}_0$ meaning that the online phase has been performed correctly.

Sub-tensor	0	1	2	3	α
True rank	3	4	5	5	—
NN-CPD	3	4	5	10	0.1
OSNCPD & OOSCPD	3	4	5	5	0.5

Table 3.
Rank estimation in the no overlapping case.

	Algorithm 1 (OSNCPD)				**Algorithm 2 (OOSCPD)**			
Sub-tensor	**0**	**1**	**2**	**3**	**0**	**1**	**2**	**3**
Mean E_A	0.17	0.19	0.19	0.2	0.17	0.18	0.2	0.16
Mean E_B	0.14	0.15	0.15	0.16	0.14	0.16	0.14	0.14
Mean E_C	0.25	0.25	0.16	0.20	0.25	0.21	0.16	0.18

Table 4.
Mean estimations errors in the overlapping case.

	Algorithm 1 (OSNCPD)				**Algorithm 2 (OOSCPD)**			
Sub-tensor	**0**	**1**	**2**	**3**	**0**	**1**	**2**	**3**
Mean E_A	0.17	0.19	0.20	0.20	0.17	0.18	0.21	0.21
Mean E_B	0.14	0.15	0.16	0.19	0.14	0.16	0.19	0.23
Mean E_C	0.25	0.32	0.31	0.32	0.25	0.31	0.32	0.25

Table 5.
Mean estimations errors in the no overlapping case.

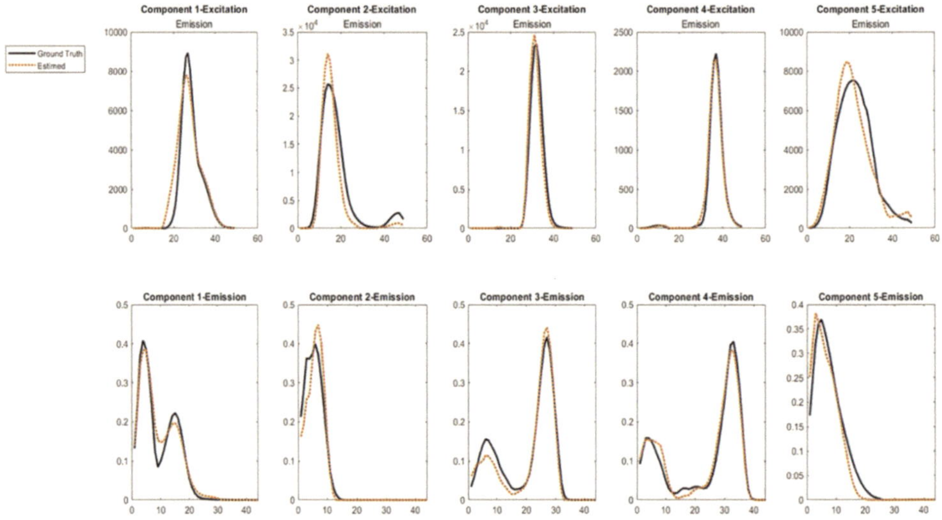

Figure 4.
Emission and excitation and spectra of the fluorophores for $\mathcal{T}_4$. Top: Emission spectra, bottom: Excitation spectra. Red dots: estimated spectra, black lines: true spectra.

In the no overlapping case OSNCPD and OOSCPD give good results although no sub-tensors are share any information. The error in matrix C seems to be high but in the no overlapping case, because the algorithms do not have the a priori information.

5. Conclusions

We have introduced two algorithms (OSNCPD and OOSCPD) for the online nonnegative canonical polyadic decomposition of sub-tensors. We have also introduced an offline nonnegative algorithm with sparsity constraint. Our algorithms are based on dictionary learning and can deal with unknown rank variations. OOSCPD incorporates orthogonality constraint and can be seen as an extension of OSNNCPD which incorporates nonnegativity and sparsity constraint for a better convergence.

These algorithms are presented in the particular case of the nonnegative online CPD of third-order fluorescence tensors, but they are not limited to this application field, and they can be easily extended to higher-order tensors. A real online fluorescence spectroscopy experiment was conducted in laboratory to validate our approach and compare with state-of-the-art approaches. The proposed algorithms allow to correctly follow the rank variations in most of the considered situations contrary to reference approaches. Eventually, these encouraging results allow to plan, for example, a monitoring in the natural environment in future work.

Conflict of interest

"The authors declare no conflict of interest."

Abbreviations

ALS	alternating least square
ADMM	alternating direction method of multipliers
CPD	canonical polyadique decomposition
NNCPD	nonnegative canonical polyadique decomposition
ELS	enhenced line search
PARAFAC	parallel factor analysis
DOM	dissolve organisme matter
EEM	emission and excitation matrices
pH	potential Hydrogen
NMF	nonnegative matrix factorization
SGD	stochastique gradient descent
Nadam	nesterov accelerated adaptive moment estimation
F	fluorescence
TRP	tryptophane
5S8HQ	5-sulfate-8-hydroxyquinine
RHO	RHOdamine
SVD	singular value decomposition

Appendix A

It is useful to mention that the CPD can incorporate different constraints depending on the field from which the data comes from. For example, it can be constraints of nonnegativity, orthogonality, or sparsity to name only the most commonly used.

Nonnegativity constraint Nonnegativity constraint is imposed in the CPD when the data to be processed is linked to physical quantities. In this sense, we can cite the domain of imagery [46] where nonnegativity is present because we know that optical signals are positive or zero. This is also the case of fluorescence spectroscopy where the data from the sensor is nonnegative [4, 30, 31, 47]. Nonnegativity constraint can be imposed on the factors during optimization by projecting the values of each factor matrix on $\mathbb{R}^+$ [33]. This method of imposing nonnegativity sometimes shows its limits in certain cases because of the change of scale. In [34], the authors propose to remedy this problem by involving two factors, ϵ and $\theta \in \{-1; 1\}$ which multiply every factor matrix. Nonnégativity can also be imposed by choosing an optimization method [14, 17, 35] or even using an exponential variable change [48] or square [31].

Sparsity constraint Regarding the constraint of parsimony, it is generally used for parsimonious tensors or to ensure the overestimation of the rank, as mentioned above. This constraint is imposed according to the goal we are looking for, and in our case, it is the overestimation of the rank. The constraint can be applied to the matrices factors in the form of a regularization term:

$$\min_x \underbrace{f(x)}_{\text{function}} + \underbrace{r(x)}_{\text{regularization}} \tag{13}$$

$F(x)$ corresponds here to $\left\| \mathcal{T} - \widehat{\mathcal{T}}(\mathbf{a}, \mathbf{b}, \mathbf{c}) \right\|_F^2$ and is the attachment part. $R(x)$ is the regularization part on the matrix **a, b,** and **c**.

In the literature, the standard $l_{1,1}$ is the most used as in [17, 35, 41, 46, 49], and it can be defined according to the Eq. (A1) as following:

$$\min_{\mathbf{a},\mathbf{b},\mathbf{c}} \left\| \mathcal{T} - \widehat{\mathcal{T}}(\mathbf{a},\mathbf{b},\mathbf{c}) \right\|_F^2 + \|\mathbf{a}\|_{1,1} + \|\mathbf{B}\|_{1,1} + \|\mathbf{C}\|_{1,1} \tag{14}$$

Standard $l_{1,1}$ is conventionally used compared to standard $l_{0,0}$ to promote parsimony because it is easier to handle than standard $l_{0,0}$. In order to minimize the rank of a matrix, we can speak of the nuclear standard which makes it possible to impose a small rank [50–52] and can be defined as a function of the Eq. (A1) as follows:

$$min_{\mathbf{a},\mathbf{b},\mathbf{c}} \left\| \mathcal{T} - \widehat{\mathcal{T}}(\mathbf{a},\mathbf{b},\mathbf{c}) \right\|_F^2 + \|\mathbf{a}\|_* + \|\mathbf{B}\|_* + \|\mathbf{C}\|_* \tag{15}$$

In [53], the authors show that the mixed standard provides a better convex envelope of the rank function than the nuclear standard. It therefore makes it possible to minimize the rank of a matrix. In the literature, the mixed standard is increasingly used as in [54–56]. It can be defined as a function of the Eq. (A1) as follows:

$$\min_{\mathbf{a},\mathbf{b},\mathbf{c}} \left\| \mathcal{T} - \widehat{\mathcal{T}}(\mathbf{a},\mathbf{b},\mathbf{c}) \right\|_F^2 + \|\mathbf{a}\|_{2,1} + \|\mathbf{B}\|_{2,1} + \|\mathbf{C}\|_{2,1} + \|\mathbf{a}\|_{1,2} + \|\mathbf{b}\|_{1,2} + \|\mathbf{c}\|_{1,2} \tag{16}$$

Appendix B: Stochastique gradient descent (SGD)

Take a differentiable function $f(\mathbf{x})$. For each iteration k, the update of $\mathbf{x}$ in gradient descent is given by

$$\mathbf{x}(k+1) = \mathbf{x}(k) - \gamma \mathbf{f}'(\mathbf{x}(k)) \tag{17}$$

with $\gamma > 0$, the step-size or the learning rate.

B.1 Adaptive moment estimation (ADAM)

Adaptive Moment estimation (ADAM) [43]. Adam optimizer is a variant of SGD. It acts on the gradient component using the exponential mobile average of the gradients and the step-size component by dividing the step-size by the square root of $\mathbf{v}$, the square of exponential average gradients.

Let take a function $f(\mathbf{x})$. For each iteration k, the update of $\mathbf{x}$ in gradient descent is given by

$$\begin{aligned} \mathbf{g}_k &= \mathbf{f}'(\mathbf{x}(k)) \\ \mathbf{m}_k &= \beta_1 \mathbf{m}_{k-1} + (1+\beta_1)\mathbf{g}_k \\ \mathbf{v}_k &= \beta_2 \mathbf{v}_{k-1} + (1+\beta_2)\mathbf{g}_k^2 \\ \hat{\mathbf{m}}_k &= \frac{\mathbf{m}_k}{1-(\beta_1)^k} \\ \hat{\mathbf{v}}_k &= \frac{\mathbf{v}_k}{1-(\beta_2)^k} \\ \mathbf{x}(k+1) &= \mathbf{x}(k) - \frac{\gamma}{\sqrt{\hat{\mathbf{v}}_k + \epsilon}} \hat{\mathbf{m}}_k \end{aligned} \tag{18}$$

with $\gamma > 0$ the step-size, $\beta_1 \in [0,1], \beta_2 \in [0,1]$ and $\mathbf{m}$, $\mathbf{v}$ is initialized to 0.

B.2 Nesterov accelerated adaptive moment estimation (Nadam)

Nadam optimizer is an extension of the Adam optimization algorithm. The algorithm was described in the 2016 article by Timothy Dozat [42]. For each iteration k of a function $f(\mathbf{x})$, the update of $\mathbf{x}$ in gradient descent is given by:

$$\begin{aligned}
\mathbf{g}_k &= \mathbf{f}'(\mathbf{x}(k-1)) \\
\mathbf{m}_k &= \beta_1 \mathbf{m}_{k-1} + (1-\beta_1)\mathbf{g}_k \\
\mathbf{n}_k &= \beta_2 \mathbf{n}_{k-1} + (1-\beta_2)\mathbf{g}_k^2 \\
\hat{\mathbf{m}}_k &= \frac{\beta_1 \mathbf{m}_k}{1-(\beta_1)^k} + \frac{(1-\beta_1)\mathbf{g}_k}{1-(\beta_1)^k} \\
\hat{\mathbf{n}}_k &= \frac{\beta_2 \mathbf{n}_k}{1-\beta_2} \\
\mathbf{x}(k+1) &= \mathbf{x}(k) - \frac{\gamma \hat{\mathbf{m}}_k}{\sqrt{\hat{\mathbf{n}}_k} + \epsilon}
\end{aligned} \tag{19}$$

Author details

Isaac Wilfried Sanou[1*†], Xavier Luciani[2†], Roland Redon[3†] and Stéphane Mounier[3†]

1 Université de Bourgogne France-comté, Bourgogne, France

2 CNRS, LIS, Université de Toulon, Aix-Marseille Université, Toulon, France

3 CNRS/INSU, Université de Toulon, Aix-Marseille Université, Toulon, France

*Address all correspondence to: isaac-wilfried.sanou@u-bourgogne.fr

† These authors contributed equally.

References

[1] Harshman RA. Foundation of PARAFAC procedure: Models and conditions for an 'explanatory' multi-mode factor analysis. In: UCLA Working Papers in Phonetics. 1970. pp. 1-84

[2] Wang H, Ahuja N. Compact representation of multidimensional data using tensor rank-one decomposition. Vectors. 2004;**1**:5

[3] Field AS, Graupe D. Topographic component (parallel factor) analysis of multichannel evoked potentials: Practical issues in trilinear spatiotemporal decomposition. Brain Topography. 1991; **3**:407-423

[4] Bro R. PARAFAC. Tutorial and applications. Chemometrics and Intelligent Laboratory Systems. 1997;**38**: 149-171

[5] Nion D, De Lathauwer L. An enhanced line search scheme for complex-valued tensor decompositions. application in DS-CDMA. Signal Processing. 2008;**88**:749-755

[6] Vasilescu MAO, Terzopoulos D. Multilinear analysis of image ensembles: Tensorfaces. In: European Conference on Computer Vision. Springer; 2002. pp. 447-460

[7] Kopriva I, Cichocki A. Blind multispectral image decomposition by 3d nonnegative tensor factorization. Optics Letters. 2009;**34**:2210-2212

[8] Sanchez E, Kowalski BR. Tensorial resolution: A direct trilinear decomposition. Journal of Chemometrics. 1990;**4**:29-45

[9] Roemer F, Haardt M. A semi-algebraic framework for approximate CP decompositions via simultaneous matrix diagonalizations (SECSI). Signal Processing. 2013;**93**:2722-2738

[10] Luciani X, Albera L. Canonical polyadic decomposition based on joint eigenvalue decomposition. Chemometrics and Intelligent Laboratory Systems. 2014;**132**:152-167

[11] Cichocki A, Zdunek R, Phan AH, Amari SI. Nonnegative Matrix and Tensor Factorizations: Applications to Exploratory Multi-way Data Analysis and Blind Source Separation. John Wiley & Sons; 2009

[12] Fu X, Huang K, Sidiropoulos ND, Ma WK. Nonnegative matrix factorization for signal and data analytics: Identifiability, algorithms, and applications. IEEE Signal Processing Magazine. 2019;**36**:59-80

[13] Phan AH, Cichocki A. Multi-way nonnegative tensor factorization using fast hierarchical alternating least squares algorithm (HALS). In: Proc. of The 2008 International Symposium on Nonlinear Theory and its Applications. 2008

[14] Liavas AP, Sidiropoulos ND. Parallel algorithms for constrained tensor factorization via alternating direction method of multipliers. IEEE Transactions on Signal Processing. 2015; **63**:5450-5463

[15] Kolda TG, Bader BW. Tensor decompositions and applications. SIAM Review. 2009;**51**:455-500

[16] Tomasi G, Bro R. A comparison of algorithms for fitting the PARAFAC model. Computational Statistics & Data Analysis. 2006;**50**:1700-1734

[17] Xu Y, Yin W. A block coordinate descent method for regularized

multiconvex optimization with applications to nonnegative tensor factorization and completion. SIAM Journal on Imaging Sciences. 2013;**6**: 1758-1789

[18] Sanou IW, Redon R, Luciani X, Mounier S. Online nonnegative canonical polyadic decomposition: Algorithms and application. In: 2021 29th European Signal Processing Conference (EUSIPCO). Vol. 2021. IEEE. pp. 1805-1809

[19] André R, Luciani X, Albera L, Moreau E. A two-step algorithm for joint eigenvalue decomposition-application to canonical polyadic decomposition of fluorescence spectra. Chemometrics and Intelligent Laboratory Systems. 2020; **206**:104065

[20] Bro R, Kiers HA. A new efficient method for determining the number of components in PARAFAC models. Journal of Chemometrics. 2003;**17**: 274-286

[21] Stedmon CA, Bro R. Characterizing dissolved organic matter fluorescence with parallel factor analysis: A tutorial. Limnology and Oceanography: Methods. 2008;**6**:572-579

[22] Papalexakis EE. Automatic unsupervised tensor mining with quality assessment. In: Proceedings of the 2016 SIAM International Conference on Data Mining, SIAM. 2016. pp. 711-719

[23] Nion D, Sidiropoulos ND. Adaptive algorithms to track the PARAFAC decomposition of a third-order tensor. IEEE Transactions on Signal Processing. 2009;**57**:2299-2310

[24] Phan AH, Cichocki A. PARAFAC algorithms for large-scale problems. Neurocomputing. 2011;**74**: 1970-1984

[25] Zhou S, Vinh NX, Bailey J, Jia Y, Davidson I. Accelerating online cp decompositions for higher order tensors. In: Proceedings of the 22nd ACM SIGKDD International Conference on Knowledge Discovery and Data Mining, ACM. 2016. pp. 1375-1384

[26] Traoré A, Berar M, Rakotomamonjy A. Online multimodal dictionary learning. Neurocomputing. 2019;**368**: 163-179

[27] Sanou IW, Redon R, Luciani X, Mounier S. Online nonnegative and sparse canonical polyadic decomposition of fluorescence tensors. Chemometrics and Intelligent Laboratory Systems. 2022;**225**:104550

[28] Tibshirani R. Regression shrinkage and selection via the lasso. Journal of the Royal Statistical Society: Series B (Methodological). 1996;**58**:267-288

[29] Mounier S, Redon R. The use of 3-d fluorescence and its decomposition in environmental organic matter studies. In: Encyclopedia of Analytical Chemistry: Applications, Theory and Instrumentation. 2006. pp. 1-16

[30] Murphy KR, Stedmon CA, Graeber D, Bro R. Fluorescence spectroscopy and multi-way techniques. PARAFAC Analytical Methods. 2013;**5**:6557-6566

[31] Royer JP, Thirion-Moreau N, Comon P, Redon R, Mounier S. A regularized nonnegative canonical polyadic decomposition algorithm with preprocessing for 3D fluorescence spectroscopy. Journal of Chemometrics. 2015;**29**:253-265

[32] Lakowicz J.. Principles of Fluorescence Spectroscopy. 2006. Vol. 1. ISBN: 978-0-387-31278-1. doi:10.1007/978-0-387-46312-4

[33] Huang K, Sidiropoulos ND, Liavas AP. A flexible and efficient algorithmic framework for constrained matrix and tensor factorization. IEEE Transactions on Signal Processing. 2016;**64**:5052-5065

[34] Jouni M, Mura MD, Comon P. Some issues in computing the cp decomposition of nonnegative tensors. In: International Conference on Latent Variable Analysis and Signal Separation. Springer; 2018. pp. 57-66

[35] Vu X, Chaux C, Thirion-Moreau N, Maire S, Carstea EM. A new penalized nonnegative third-order tensor decomposition using a block coordinate proximal gradient approach: Application to 3d fluorescence spectroscopy. Journal of Chemometrics. 2017;**31**:2859

[36] Royer JP, Thirion-Moreau N, Comon P. Computing the polyadic decomposition of nonnegative third order tensors. Signal Processing. 2011;**91**: 2159-2171

[37] Vu, X.T., Chaux, C., Maire, S., Thirion-Moreau, N., 2014. Study of different strategies for the canonical polyadic decomposition of nonnegative third order tensors with application to the separation of spectra in 3d fluorescence spectroscopy. In: 2014 IEEE International Workshop on Machine Learning for Signal Processing (MLSP), IEEE. pp. 1–6.

[38] Kruskal JB. Three-way arrays: Rank and uniqueness of trilinear decompositions, with application to arithmetic complexity and statistics. Linear Algebra and its Applications. 1977;**18**:95-138

[39] Pasricha R, Gujral E, Papalexakis EE. Identifying and alleviating concept drift in streaming tensor decomposition. In: Joint European Conference on Machine Learning and Knowledge Discovery in Databases. Springer; 2018. pp. 327-343

[40] Cohen JE, Gillis N. Dictionary-based tensor canonical polyadic decomposition. IEEE Transactions on Signal Processing. 2017;**66**:1876-1889

[41] Lee H, Battle A, Raina R, Ng AY. Efficient sparse coding algorithms. In: Advances in Neural Information Processing Systems. Citeseer; 2007. pp. 801-808

[42] Dozat T. Incorporating Nesterov momentum into Adam. In: Proceedings of 4th International Conference on Learning Representations, Workshop, Track. 2016

[43] Kingma DP, Ba J. Adam: A method for stochastic optimization. arXiv preprint arXiv:1412.6980. 2014

[44] Zdunek R, Cichocki A. Fast nonnegative matrix factorization algorithms using projected gradient approaches for large-scale problems. Computational Intelligence and Neuroscience. 2008, 2008

[45] Olikier G, Absil PA, De Lathauwer L. Variable projection applied to block term decomposition of higher-order tensors. In: International Conference on Latent Variable Analysis and Signal Separation. Springer; 2018. pp. 139-148

[46] Zhang Q, Wang H, Plemmons RJ, Pauca VP. Tensor methods for hyperspectral data analysis: A space object material identification study. Journal of the Optical Society of America A. 2008;**25**:3001-3012

[47] Coble PG, Lead J, Baker A, Reynolds DM, Spencer RG. Aquatic Organic Matter Fluorescence. Cambridge University Press; 2014

[48] Royer JP. Identification aveugle de mélanges et décomposition canonique de tenseurs: Application à l'analyse de l'eau. [theses]. Université Nice Sophia Antipolis. 2013

[49] Cohen JE, Gillis N. Dictionary-based tensor canonical polyadic decomposition. IEEE Transactions on Signal Processing. 2018;**66**:1876-1889. DOI: 10.1109/tsp.2017.2777393

[50] Gu S, Zhang L, Zuo W, Feng X. Weighted nuclear norm minimization with application to image denoising. In: Proceedings of the IEEE Conference on Computer Vision and Pattern Recognition. 2014. pp. 2862-2869

[51] Recht B, Fazel M, Parrilo PA. Guaranteed minimum-rank solutions of linear matrix equations via nuclear norm minimization. SIAM Review. 2010;**52**: 471-501

[52] Yuan M, Zhang CH. On tensor completion via nuclear norm minimization. Foundations of Computational Mathematics. 2016;**16**: 1031-1068

[53] Han X, Albera L, Kachenoura A, Senhadji L, Shu H. Low rank canonical polyadic decomposition of tensors based on group sparsity. In: 2017 25th European Signal Processing Conference (EUSIPCO). IEEE; 2017. pp. 668-672

[54] Chen Y, He W, Yokoya N, Huang TZ. Hyperspectral image restoration using weighted group sparsity-regularized low-rank tensor decomposition. IEEE Transactions on Cybernetics. 2019;**50**:3556-3570

[55] Gramfort A, Kowalski M, Hämäläinen M. Mixed-norm estimates for the M/EEG inverse problem using accelerated gradient methods. Physics in Medicine & Biology. 2012;**57**:1937

[56] de Morais Goulart JH, de Oliveira PMR, Farias RC, Zarzoso V, Comon P. Alternating group lasso for block-term tensor decomposition and application to ECG source separation. IEEE Transactions on Signal Processing. 2020; **68**:2682-2696

Chapter 4

The Combined Search of the Different Kind of Solutions to the Optimization Problems of Antenna Synthesis Theory

Mykhaylo Andriychuk

Abstract

The restriction on the optimized function in the problem of the antenna synthesis leads to the characteristic form of the nonlinear integral equations, which are equivalent to the respective functionals of synthesis problem. The kind of the reduced equation depends on the operator, which is used for the calculation of antenna's directivity pattern (DP). The two types of restrictions are considered, the first one is that the amplitude of the objective function is prescribed. The second one is that the restrictions on the phase of the objective function are imposed. The problem is actual because it deals with the analytical investigation of integral Hammerstein equations of a new type, on the one hand, and gives a possibility to get new types of solutions to above equations that can be applied at the modelling of antennas of the different appointment, on the other hand. The peculiarity of such equations is the non-uniqueness of their solutions. The accuracy of analytical approach is supported by the numerical data. The method of successive approximations is much effective for solving the reduced equations. The advantage of the approach proposed is a fast convergence of iterative procedures, which are applied for solving the obtained equations.

Keywords: amplitude radiation pattern, variational approach, restriction on the objective function, non-linear equations, branching of solutions, methods of successive approximation, modeling results

1. Introduction

Optimization with restrictions is an effective method for solving the different application problems in the various fields of engineering and technology [1–4]. The problems discussed therein appertain to both the general schemes of algorithm evolution [1, 2] and using the optimization algorithms for solving the specific problems related to engineering and technology applications [3, 4]. The numerous papers cited therein testify that the majority of optimization problems appearing in the engineering design are related to optimization problems with restrictions.

The antenna synthesis theory and technique [5] represent one of the above engineering fields. Usually, while solving the problem of synthesis for the different types of antenna systems, the restrictions on the physical parameters and parameters of radiation are applied. The typical synthesis problem of antenna is reduced to creating the current or field distribution in antenna construction, which produces the prescribed radiation data. In our consideration, the amplitude directivity pattern (DP) is considered as such feature [6]; there various problems of optimization belonging to the different types of antenna systems were examined. The optimization problem is stated in the variational form because the given amplitude DP can not be formed exactly. Such statement provides the ability to create the amplitude DP of antenna in the closest form to what is desired.

The amplitude DP is not the only one given characteristic for optimization in the antenna synthesis problems. The majority of radiation parameters and characteristics are subject to the interest. One is required to take into account the limitations to the antenna construction in many cases [7, 8], the electromagnetic compatibility demands [9–11], the peculiarities of whole radiation [12–14], and the level of DP's sidelobes [15, 16]. All the mentioned problems can be formulated as the problems of multiparameter optimization; therefore it is suitable to apply the variational formulation for their statement and further solving. The above is an additional reason to use a variational approach for solving the optimization problem we considered in this chapter.

The variational formulation the antenna synthesis optimization problems was proposed for the first time in [17]; as a part of this research, the theoretical base, which enables to get a portion of information regarding the solutions of problems, was developed later in [5]. Since not complex DP of the antenna is optimizing function, but only its amplitude, the variational formulation of the synthesis problem by the amplitude DP leads to the need to solve the non-linear integral equations.

The resulting equations are the Hammerstein non-linear integral equations. The non-uniqueness and process of branching their solutions is the peculiar property of such equation [18]. Usually, only one unique solution exists when a characteristic parameter in a kernel of such equations is small; and a solution becomes branching when this parameter grows. In fact, we need to use the numerical methods to solve such equations, because the solution cannot be derived analytically. A series of such methods was developed in [17]. The above method is characterized by that we start with an initial approximation having some property of the phase of DP, and the resulting solution is determined in class of functions, which have the same property of phase DP.

This approach was generalized in [5], the solutions in a wide class of functions with argument (or phase) in form of a finite degree polynomial, was sought for there. Employment of the Newton generalized method, allowing simultaneous search for some branched solution and determination the characteristic parameter, at which the solution becomes branching, is one more excellence of this approach. The proposed approach correlates with the variety of Newton methods developed in [19]; and it can be effectively applied to solve a series of engineering problems for antennas of different kinds. The developed methods were applied in [6] at solving the synthesis problems of microstrip antennas, planar arrays, and antennas with waveguide excitation. The initial requirement while solving the problem of synthesis knows the recipe how the antenna's DP should be calculated. The method proposed in [20], is applied to derive the closed form of the DP.

The specific of the problems solved in works [5, 6, 17] was either the absence of restrictions on the function to be sought for (the synthesis by the amplitude and phase), or the restrictions are applied to the amplitude of this function (the synthesis by phase), and the restrictions are applied to the phase of this function (the synthesis by amplitude). The problems, which are solved here, belong to the phase and amplitude synthesis in the context of the problems studied in [5, 17, 21].

This chapter is arranged as follows. In Sections 2 and 3, we formulate the problems of phase and amplitude optimization. The non-linear integral equations of the used functionals are reduced here in the general operator form. The specification of such operators for the 1D and 2D cases is given. Some aspects of convergence of the successive approximation method and properties of obtained non-linear equations are discussed as well. The data of numerical modeling presented in Section 4 are related to the different types of antennas and demonstrate the peculiarities of the solutions to respective non-linear equations and rate of convergence of the successive approximation method, which is used.

The concluding remarks finish the considerations and accentuate the engineering importance of the approach proposed.

2. Phase optimization

2.1 The optimization problem outlines

In this Section, we deal with the optimization problem, which is characterized by restrictions on the amplitude of the function is sought for. Such optimization problem is related to solving the phase synthesis of antenna systems corresponding to the linear antennas and arrays. The function to be optimized presents some vector, which describes the phase value of currents I in the elements of the array. The constrains on the amplitudes $|I|$ of currents are characterized by prescribing some determined function. A non-negative function F is the input data of problem; it is approximated with the amplitude of the created (synthesized) DPf. The DPf is determined by some linear bounded operator A using the currents I in the antenna elements.

For our approach, we give $|I| = C|v|$, where the value of $|v|$ is fixed, and C unknown real number to be determined below. The functional, which described an optimization problem, is [5].

$$\sigma(\psi, C) = \left\| C|A(|v|e^{i\psi})| - F \right\|^2, \tag{1}$$

where $\|\cdot\|$ is a norm of vectors, prescribed in form of inner product in the Hilbertian space of functions, to which the DPs belong.

The number C is determined by some relation; i.e. it depends on a phase function ψ in such a way:

$$C(\psi) = \left(A\left(|v|e^{i\psi}\right), F\right) / \left\|A\left(|v|e^{i\psi}\right)\right\|^2 \tag{2}$$

Using (1) and (2) one can get that $\sigma(\psi, C) = \|F\|^2 - \kappa(\psi)$; the last relation associates $\sigma(\psi, C)$ with functional

$$\kappa(\psi) = \frac{\left(|A\left(|v|e^{i\psi}\right)|, F\right)}{\left\|A\left(|v|e^{i\psi}\right)\right\|} \tag{3}$$

One can easy to sure that (3) does not depend on the number C. The function-vector ψ, which is looking for, maximizes $\kappa(\psi)$ functional; simultaneously, it gives a minimum for functional (1), when the number C is determined according to (2). Therefore, one can search for a minimum of functional (1) or a maximum of functional (3) when the optimizing function ψ is determined as solution to the phase synthesis problem. I turned out that in contrast to the problem of amplitude-phase synthesis [6], the problem of maximization of functional (3) is more convenient while solving the phase synthesis problem than minimization of functional (1); the above is result of that that one can establish a constructive iterative process, which provides a quick convergence in phase optimization.

2.2 Reducing the non-linear equation

In this subsection, we reduce the Lagrange–Euler equation associated with functional (3). To do this, we add to the function ψ term $\delta\psi$, where δ is an unknown small parameter (real number), and present the increment in the form

$$v + \delta v = |v|e^{i(\psi+\delta\psi)} = |v|e^{i\psi}(1 + i\delta\psi) \tag{4}$$

up to first-order terms.

The above results in

$$\delta v = ie^{i\psi}i\delta\psi \tag{5}$$

and

$$\delta f = iA\left(|v|e^{i\psi}i\delta\psi\right) \tag{6}$$

The last formula takes into account the operator A is linear, and $f = A\left(|v|e^{i\psi}\right)$. If to take into account (5) and (6), we reduce the incremental part of functional (3)

$$\delta\kappa(\delta\psi) = \left(\delta\psi, |v|\mathrm{Im}\left(e^{-i\psi}A^*\left[Fe^{i\arg f} - Cf\right]\right)\right) \tag{7}$$

and A^* is an adjoint operator to A.

2.3 The non-linear Lagrange: Euler equation in the generalized form

To reduce the Lagrange–Euler equation for the unknown function ψ, we derive the first variation of functional (3). This variation has form

$$\begin{aligned}
&\delta\kappa(\psi, \delta\psi) = \mathrm{Im}(A[|v|\exp(i\psi)\delta\psi], F\exp(i\arg f) - Cf) = \\
&\mathrm{Im}(|v|\exp(i\psi)\delta\psi, A^*[F\exp(i\arg f) - Cf]) = \\
&(\delta\psi, |v|\mathrm{Im}(\exp(-i\psi)A^*[F\exp(i\arg f) - Cf])).
\end{aligned} \tag{8}$$

The condition

$$\mathrm{Im}\left(e^{-i\psi}A^*\left[Fe^{i\arg f} - Cf\right]\right) = 0 \tag{9}$$

must be met to satisfy the necessary condition of extreme of $\kappa(\psi)$. Eq. (9) is governing the Lagrange–Euler equation, which allows to search for the unknown distribution ψ of phase.

One can reduce two relating equations using (9); they are:

$$\psi = \arg\left(A^*\left[Fe^{i\arg f} - Cf\right]\right) \tag{10}$$

and

$$\psi = \arg\left(A^*\left[Cf - Fe^{i\arg f}\right]\right) \tag{11}$$

The last derived equations are equivalent; in this connection, the first one we use for the numerical calculations.

We should emphasize that (10) is non-linear equation with respect to the unknown function ψ, the above is result that ψ contained implicitly in the exponent of function $i\arg f$, and additionally $f = A\left(|v|e^{i\psi}\right)$.

2.4 The simplified equation of Lagrange-Euler

In the process of numerical experiments, more simple functional

$$\kappa_s(\psi) = \left(\left|A\left(|v|e^{i\psi}\right)\right|, F\right) \tag{12}$$

which is the numerator of functional (3), is applied at solving the practical problems. The respective to (12), the Lagrange–Euler equation is

$$\mathrm{Im}\left(e^{-i\psi}A^*\left[Fe^{i\arg A\left(|\mathbf{v}|e^{i\psi}\right)}\right]\right) = 0 \tag{13}$$

and (13) provide us with an explicit formula

$$\psi = \arg\left(A^*\left[Fe^{i\arg f}\right]\right) \tag{14}$$

for determination of the unknown distributions of the phases of currents. Eqs. (13) and (14) are more suitable for the numerical study than Eqs. (9) and (10), because they are free of an auxiliary number C.

On the other hand, we can consider the additional function

$$w = A^*\left[Fe^{i\arg f}\right] \tag{15}$$

that gives $\psi = \arg w$, and

$$f = A^*\left[Fe^{i\arg w}\right] \tag{16}$$

Eqs. (15) and (16) are equivalent to Eq. (14). They present non-linear system of equations for two unknown functions w and f.

The reduced forms (13), (15), and (16) of non-linear Lagrange–Euler equations results in the subsequent iterative process to determine their solutions

$$\psi^{(p+1)} = \arg\left(A^*\left[Fe^{i\arg f^{(p)}}\right]\right) \tag{17}$$

for (14) and

$$f^{(p)} = A^*\left[Fe^{i\arg w^{(p)}}\right] \tag{18}$$

$$w^{(p+1)} = A^*\left[Fe^{i\arg f^{(p)}}\right] \tag{19}$$

for (15) and (16), respectively. The convergence of the obtained iterative processes was studied firstly in [6]. Eqs. (17)–(19) are non-linear, therefore the peculiarities of their solutions depend on the properties of the A operator, the frequency of radiation, and the geometry of array.

2.5 The operators A and A^* in the 1D and 2D cases

One can conclude from Eqs. (18)–(19) that the iterative process of its solving needs many actions of the operators A and A^* on the interim unknown functions. As a rule, the operator A corresponds to calculation of the DP f of array by the currents I on its radiators. The form of operator A^* is defined by the geometry of array and the used inner products in the spaces of DPs and currents [17].

2.5.1 The operators A and A^* in the 1D case

If we solve the synthesis problem in the 1D case, the above is related to a linear array, operator A determines the DP of array and acts as

$$f(\xi) = AI = \sqrt{\frac{c}{2\pi}} \sum_{n=-N}^{N} I_n \exp(icn\xi) \tag{20}$$

i.e., operator A correspons to the discrete Fourier transform, and it acts from the Euclidean complex $(2N+1)$–dimensional space $l_2(-N, N)$ to Hilbertian space $L_2(-\pi/c, \pi/c)$ of continuous functions. In Eq. (20), parameter ξ is a generalized angular coordinate; it is determined by relation $\xi = \sin\theta / \sin\theta_0$, and value $2\theta_0$ is angle, where the function F is not equal to zero; $-\pi/2 \le \theta \le \pi/2$. The DP (20) is periodical and its period is equal to $2\pi/c$.

The operator A in formula (20) is isometric [5]. The operator A^* acts from space $L_2(-\pi/c, \pi/c)$ into space $l_2(-N, N)$, it has form

$$A^*v = \sqrt{\frac{c}{2\pi}} \int_{-\pi/c}^{\pi/c} v(\xi) \exp(-icn\xi) d\xi, \; n = -N, \dots, N \tag{21}$$

We use also the operator $AA^*: l_2(-N, N) \to l_2(-N, N)$

$$[AA^*v](\xi) = \frac{c}{2\pi} \int_{-\pi/c}^{\pi/c} v(\xi') \frac{\sin[c(N+1/2)(\xi - \xi')]}{\sin[c(\xi - \xi')/2]} d\xi' \tag{22}$$

which corresponds to kernel of the integral non-linear equation

$$Aw = A\left[A^*\left(Fe^{i\arg f}\right)\right] \tag{23}$$

and which is solved to determine the DP f. The last equation is applied mainly while the analytical research regarded to the non-uniqueness of solution and the process of branching [5].

2.5.2 The operators A and A^* in the 2D case

The application of the method developed here is used to a plane equidistant array. We consider an array having the odd number of radiators in the direction of both coordinate axes, and take into account only array factor but not whole DP [11]. Additionally, we use in a far zone the generalized angular coordinate. The above assumptions allow us to consider the DP of array as a product of independent DPs in the $O'\xi_1$ and $O'\xi_2$ planes that differs on use of usual coordinates θ and φ. We assume also that the current's phases in the array's radiators are separated on two independent distribution sets along the axes Ox and Oy, and coordinate system $\xi_1 O'\xi_2$ corresponds to a specific set in a far zone. When to take into account the above assumptions, we present the whole DP in form

$$f(\xi_1, \xi_2) = A\mathbf{I} \equiv \sum_{m=-M_1}^{M_1} \sum_{n=-N_1}^{N_1} g_{nm}(\xi_1, \xi_2) I_{nm} e^{i(c_1 n\xi_1 + c_2 m\xi_2)} \tag{24}$$

and $g_{nm}(\xi_1, \xi_2)$ is the DP of single radiator, $\mathbf{I} = \{I_{nm}\}$ are currents in the radiators of array. The new coordinates are given in form $\xi_1 = \sin\theta\cos\varphi/\sin\theta_1$, and $\xi_2 = \sin\theta\sin\varphi/\sin\theta_2$, where $k = 2\pi/\lambda$ is the wavenumber, λ is length of wave; $c_1 = kd_1\sin\theta_1$, $c_2 = kd_2\sin\theta_2$; the values d_1, d_2 correspond to the distances between separate radiators in the Ox and Oy directions, respectively. We assume also that the prescribed DP F is zero beyond $[\pm\theta_1, \pm\theta_2]$.

If the scalar products in the spaces of DPs and currents are determined according to Ref. [6], one can reduce the close form for the corresponding elements of the operator A^*

$$[A^*f]_{nm} = \frac{c_1 c_2}{4\pi^2} \int_{-1}^{1}\int_{-1}^{1} f(\xi_1, \xi_2) g_{nm\,*}(\xi_1, \xi_2) \times e^{-i(c_1 n\xi_1 + c_2 m\xi_2)}] d\xi_1 d\xi_2. \tag{25}$$

Eq. (24) is reduced by the assumption that the mutual coupling the adjacent radiators is taken into account by using the DPs $g_{nm}(\xi_1, \xi_2)$ of separate elements of array. It can be reduced to simpler form if to take this influence while solving the direct (the analysis) problem in the case of plane array [9].

The iterative processes (17) or (18)–(19) can be easy in use, when the function A^*f is known.

3. Amplitude optimization

3.1 The optimizing functional

The problem of the amplitude optimization reduces to minimization of functional

$$\sigma = \| |A[v\exp(i\psi)]| - F \|^2 \tag{26}$$

where F is real and positive function (this is desired or given amplitude DP), function ψ is phase (argument) of the complex current $u = v\exp(i\psi)$ in antenna; term $\|\cdot\|$ is used to define the norm of function in the spaces of DPs or currents. Operator A describes the definition of the DP f of antenna by known functions v and ψ, i.e. solves a direct electro-dynamical problem. The closed form of operator A is very important when the optimization problems are solved; moreover, this is essential not only in the synthesis problems but also in the problems of determining the material parameters of scattering medium [22]. The additional terms can be included into (26) to impose specific requirements to the given function F [23].

In the case is considered, phase ψ of function u is given, the minimization of (26) is carried out by choosing the function v. In functional (26), it is assumed that function v must be real but not necessary positive. Namely, it can have different sign in the points of definition. To do it positive everywhere, we can add jump equal to π in function ψ, where it is negative. Therefore, the function v is determined in a set of positive functions. The functional (26) can by minimized directly by the gradient methods [24] while the necessity to solve the specific engineering problems; and the reduction to the corresponding Euler equation is one more recipe of minimization. The latter allows to study the properties of the obtained non-linear equations and the convergence of proposed method of the successive approximations.

3.2 The Euler equation for functional

We reduce the first variation with respect to the unknown function v in functional (26) to reduce the corresponding Euler equation. For this end, we substitute function v in formula (26) by $v_0 + \delta v$ (δ is small number), and we deal with the first-order terms with respect to the increment δv. After this, we have

$$\sigma(v_0 + \delta v) = \sigma(v_0) + 2\,\mathrm{Re}(\delta f, f) - 2(\delta|f|, F), \tag{27}$$

where $(\cdot,\cdot)$ is a scalar (inner) product of the functions in DPs space, and

$$f = A[v\exp(i\psi)], \tag{28}$$

$$\delta f = A[\delta v\exp(i\psi)], \tag{29}$$

$$\delta|f| = \mathrm{Re}[\delta f\exp(-i\arg f)]. \tag{30}$$

Using two latter equalities, we derive

$$\delta v(v, \delta v) = 2(\delta v,\ \mathrm{Re}\{\exp(-i\psi)A^*(f) - \exp(-i\psi)A^*[F\exp(-i\arg f)]\}), \tag{31}$$

where A^* is operator adjoint to A [17].

The formula (31) allows us to get the Euler equation corresponding to (26)

$$\mathrm{Re}\{\exp(-i\psi)A^*A[v\exp(i\psi)] - \exp(-i\psi)A^*[F\exp(i\arg A[v\exp(i\psi)])]\} = 0. \tag{32}$$

In such a way, (32) is the Hammerstein non-linear integral equation, and the iterative process

$$\mathrm{Re}\left\{\exp(-i\psi)A^*A\left[v^{(n+1)}\exp(i\psi)\right]-\exp(-i\psi)A^*\left[F\exp\left(i\arg A\left[v^{(n)}\exp(i\psi)\right]\right)\right]\right\}=0 \tag{33}$$

is applicable to solve it.

3.3 The convergence of method

To prove the convergence of iteration process (33), we study the auxiliary functional [5].

$$\sigma_{(n)}(v)=\left\|\left|A[v\exp(i\psi)]\right|-F\exp(i\varphi_{(n)})\right\|^2 \tag{34}$$

where $\varphi_{(n)}=\arg\left(f^{(n)}\right)$, and $f^{(n)}=A\left[v^{(n)}\exp(i\psi)\right]$.

The Euler equation for the last functional has form

$$\mathrm{Re}\left\{\exp(-i\psi)A^*A[v\exp(i\psi)]-\exp(-i\psi)A^*\left[F\exp\left(i\varphi_{(n)}\right)\right]\right\}=0, \tag{35}$$

and one can prove that its solution coincides with function $v^{(n+1)}$. This fact testifies that $v^{(n+1)}$ provides a minimum for (34).

Taking into account the results of [5] (subsection 3.2, page 51), we can write

$$\sigma_{(n)}(v)-\sigma(v)=2\left(|f|\left[1-\cos\left(\arg f-\varphi_{(n)}\right)\right],F\right)\ge 0 \tag{36}$$

and the last inequality results in the chain let

$$\sigma\left(v^{(n+1)}\right)\le\sigma_{(n)}\left(v^{(n+1)}\right)\le\sigma_{(n)}\left(v^{(n)}\right)=\sigma\left(v^{(n)}\right) \tag{37}$$

which proves the convergence of iterative process (33).

The non-linear Eq. (32) has non-unique solutions, and such solutions branch off when the number c in the kernel A^*A grows. As a rule, there exists the unique solution only at small c. While necessity to find the branching points we should consider the corresponding homogeneous equation and to solve the respective eigenvalue problem.

3.4 Equation to determine the branching points

As in the case of amplitude-phase and phase synthesis problems [17], the non-linear Eq. (32) becomes the non-unique solutions if the characteristic parameter grows in its kernel A^*A (only a unique solution exists if this parameter is small). The method of perturbations is used to get a homogeneous equation for determination of the points of branching. It is evidently that perturbation of the parameter $c=c_0+\varepsilon$ in the kernel leads to perturbation of the operator

$$A=A_0+\varepsilon A_1 \tag{38}$$

and the solution

$$v = v_0 + \varepsilon v_1 \tag{39}$$

If we substitute Eqs. (38), (39) into (32), and after some transformations, we get the non-homogeneous linear equation

$$\begin{aligned}&\mathrm{Re}\left[\exp(-i\psi)A_0^*[v_1\exp(i\psi)] - iF\exp(i\arg(f_0)) \times \mathrm{Im}\frac{A_0(v_1\exp(i\psi))}{f_0}\right] = \\ &-\mathrm{Re}\left[\exp(-i\psi)A_1^*A_1[v_0\exp(i\psi)]\right] \\ &+\mathrm{Re}\left[\exp(-i\psi)A_1^*[F\exp(i\arg A_1[v_0\exp(i\psi)])]\right]\end{aligned} \tag{40}$$

for the perturbed function v_1, and

$$f_0 = A_0(v_0\exp(i\psi)) \tag{41}$$

Eq. (40) has an unique solution when the homogeneous equation

$$\mathrm{Re}\left[\exp(-i\psi)\left[A_0^*[v_1\exp(i\psi)] - iF\exp(i\arg(f_0)) \times \mathrm{Im}\frac{A_0(v_1\exp(i\psi))}{f_0}\right]\right] = 0 \tag{42}$$

does not have the nontrivial solutions. The different solutions to (32) can be exist at the parameter c, at which Eq. (42) has a nontrivial solution. Basing on this results, we get the non-linear eigenvalue problem with respect to c, and we write it in the form

$$\begin{aligned}&\lambda_n\,\mathrm{Re}\left[\exp(-i\psi)\left[A_0^*[A_0(w_n\exp(i\psi))]\right]\right] = \\ &-\mathrm{Im}\left[\exp(-i\psi)A_0^*\left[F\exp(i\arg(f_0))\mathrm{Im}\frac{A_0(w_n\exp(i\psi))}{f_0}\right]\right].\end{aligned} \tag{43}$$

The value $c = c_0$ is the point of branching for solution of Eq. (32) when at least one of its eigenvalues λ_n in (43) is unity.

3.5 The cases of different operators *A*

One can show that form of Eq. (32) can vary for the different operators A. We confine ourselves by operator A, which corresponds the DP of a plane continuous antenna (the case of compact operator), and the DP of a linear antenna (the case of isometric operator).

3.5.1 The compact operator A

For this case, the method of minimization of (26) is unstable; namely, its solutions become perturbed considerably, if the amplitude DP F or the operator A becomes small perturbation. In order to exclude this drawback, we add tot (26) some stabilization term

$$\sigma_\alpha(v) = \||A[v\exp(i\psi)]| - F\|^2 + \alpha\|v\|^2 \tag{44}$$

where number $\alpha > 0$ is the value of regularization parameter.

The operator A in this case becomes form

$$f(\varphi) = Au = \int_{-\pi}^{\pi} u(\varphi') \exp[ikr(\varphi') \cos(\varphi - \varphi')] dS_{\varphi'} \tag{45}$$

where angular coordinate φ and φ' are points in the antenna and a far zone, the wavenumber k plays role of parameter c, and element of arc is $dS_{\varphi'} = \sqrt{r^2(\varphi') + (dr(\varphi\prime)/d\varphi\prime)^2}$.

The Euler equation for (44) has form

$$\mathrm{Re}\{ \exp(-i\psi)A^* A[v\exp(i\psi)] - \exp(-i\psi)A^* [F\exp(i\arg A[v\exp(i\psi)])]\} + \alpha v = 0, \tag{46}$$

and form of the ajoint operator A^* is determined by form of operator A and the iscalar products in the spaces of the currents and DPs. We use such iterative process to solve it

$$\begin{aligned} &\mathrm{Re}\left\{ \exp(-i\psi)A^* A\left[v^{(n+1)} \exp(i\psi)\right]\right\} + \alpha v^{(n+1)} \\ &= \mathrm{Re}\left\{ \exp(-i\psi)A^* \left[F\exp\left(i\arg A\left[v^{(n)} \exp(i\psi)\right]\right)\right]\right\} = 0. \end{aligned} \tag{47}$$

The corresponding eigenvalue problem looks like similarly to (43), but it is supplemented by term $\lambda_n \alpha w_n$ in the first line.

3.5.2 The isometric operator A

The equality is true

$$\|A(v\exp(i\psi))\| = \|v\| \tag{48}$$

if operator A is an isometric. It connects the norms of the amplitude v and DP $f = A(v\exp(i\psi))$, therefore (26) becomes

$$\sigma(v) = \|v\|^2 - 2(|A[v\exp(i\psi)]|, F) + \|F\|^2. \tag{49}$$

The operator A for determination of the DP is

$$f(\xi) = Au = \int_{-1}^{1} u(x) \exp(ic\xi x) dx \tag{50}$$

In this case, the Euler equation becomes simpler form

$$v = \mathrm{Re}\{ \exp(-i\psi)A^* [F\exp(i\arg A[v\exp(i\psi)])]\} \tag{51}$$

and the iterative procedure is

$$v^{(n+1)} = \mathrm{Re}\left\{ \exp(-i\psi)A^* \left[F\exp\left(i\arg A\left[v^{(n)} \exp(i\psi)\right]\right)\right]\right\} \tag{52}$$

The points of branching of solution to Eq. (51) are that values of the number c, at which the corresponding linear homogeneous equation

$$\lambda_n w_n = -\mathrm{Im}\left\{F\exp(i\arg(f_0))\mathrm{Im}\frac{A_0[w_n\exp(i\psi)]}{f_0}\right\} \tag{53}$$

has the multiple eigenvalues $\lambda_i = 1$.

4. Numerical modeling

4.1 Phase optimization

We show in this section the results related to solving the synthesis problems for the 1D and 2D cases, which are related to the engineering implementation for the linear antennas and plane equidistant arrays [14].

In the case if A operator is Fourier transform of some limited function that corresponds to physical nature of a linear antenna, the optimization problem similar to (26) was solved firstly in [6]; the iterative method (the algorithm proposed by Katsenelenbaum and Semenov, [5]) for its solving was proposed and acquitted. In paper [17], it was borrowed to the Fourier transform in the discrete case, and a series of the numerical results was got.

The different solutions supplementing a maximum of (3), which correspond to the non-linear system (18) and (19), are presented here. Such solutions give maximum at the certain values of the characteristic number c, which define the frequency of radiation and geometry of antenna. The global maximum is defined by the upper envelope of curves match the different types of solutions.

In our case, system (18), (19) has form

$$w(x) = \int_{-1}^{1} F(\xi)\exp(i\arg f(\xi))\exp(-icx\xi)d\xi \tag{54}$$

$$f(\xi) = \int_{-1}^{1} U(x)\exp(i\arg w(x))\exp(icx\xi)dx \tag{55}$$

4.1.1 The variety of solutions

In the case if U and F are the even functions of their arguments, the solutions to Eqs. (18) and (19) can be specified within three groups:

a. the functions $w(x), f(\xi)$ are real valued for $x \in (-1, 1)$, $\xi \in (-1, 1)$, respectively; either one or both such functions are non-zero;

b. one of function, let $w(x)$, is real valued and becomes zeros in the interval $(-1, 1)$; the other function $f(\xi)$ has an odd phase, i.e. $\varphi(-\xi) = -\varphi(\xi)$;

c. the functions $f(\xi)$, $w(x)$ have an even phases, namely $\psi(-x) = \psi(x)$, $\varphi(-\xi) = \varphi(\xi)$.

The solutions (c), for which both the functions $f(\xi)$, $v(\xi)$ are complex valued, exist in the case of symmetrical and non-symmetrical functions $U(x)$, $F(\xi)$. Furthermore, the global extremal points of κ_s are located between above solutions, except the case if they both are defined by the Eqs. (26) and (27) for the symmetrical data. Taking into account the above, we consider below the solutions that are determined numerically, and the symmetry of data does not take into account.

4.1.2 The numerical solutions

We consider firstly the case, if both the functions $v(x)$, $F(\xi)$ are identical. The results for $v(x) = \text{const}$, $F(\xi) = \text{const}$ are presented below. The values of σ functional for various solutions to Eqs. (18) and (19) are shown in **Figure 1**. There exists one trivial solution $w_0(x) = 0, f_0(\xi) = 0$ up to point $c = \pi$ (this value is denoted by c_2). Nevertheless, this solution ceases as optimal earlier, i.e., at the point $c = c_1 < c_2$. This value of c (it is ≈ 2.80) corresponds to the first point, in which the real solution branches off.

The solution $\{f_0, w_0\}$ branches into four different at the point $c = c_2$. Three of such solutions are real. The first of them (denoted by 0) is symmetrical, and the functions $f_0(\xi)$, $w_0(x)$ coincide in the shape for it. The functions $\varphi_0(\xi)$ (phase of $f_0(\xi)$) and $\psi_0(x)$ (phase of $w_0(x)$) are also the same, of the form $\varphi_0(\xi) = \begin{cases} 0, f_0(\xi) \geq 0, \\ \pi, f_0(\xi) < 0. \end{cases}$. Their zeros are shown in **Figure 2** as $x_{1,2}^{(0)}, \xi_{1,2}^{(0)}$. Two remaining solutions (marked by 0 and 0′) are characterized by the different $f_{0'}(\xi)$ and $w_{0'}(x)$. They are mirror-symmetrical, i.e. these functions are interchanging in them. One of such functions (let $f_0(\xi)$) can be calculated by $f_0 = Av$; by this, its zeros are shown in **Figure 3** as $\xi_m, m = 1, 2, \dots, 5, 6$. Another function (namely $f_{0'}(\xi)$) has unit sign for the examined parameters of c.

A global maximum for this set of solutions is described by the different solutions at the defined values of c (see **Figure 1**). One should note that the solution $\{f_0, w_0)$ is the best immediately after $c = c_2$.

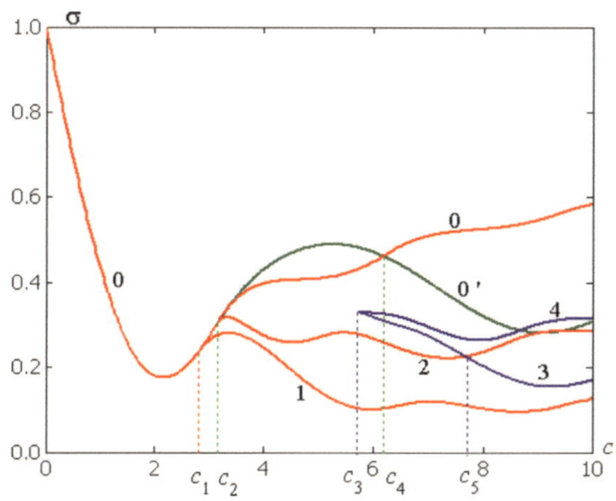

Figure 1.
The values of functionals $\sigma = 1 - |\kappa_s|^2$.

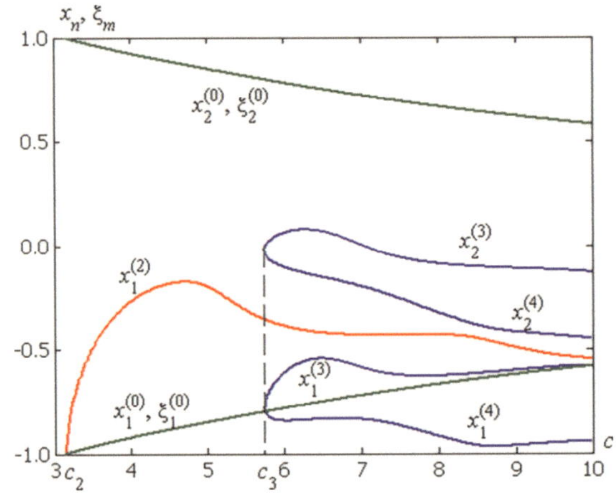

Figure 2.
Zero curves of the functions $w(x)$ and $f(\xi)$, corresponding to the real and complex symmetrical solutions.

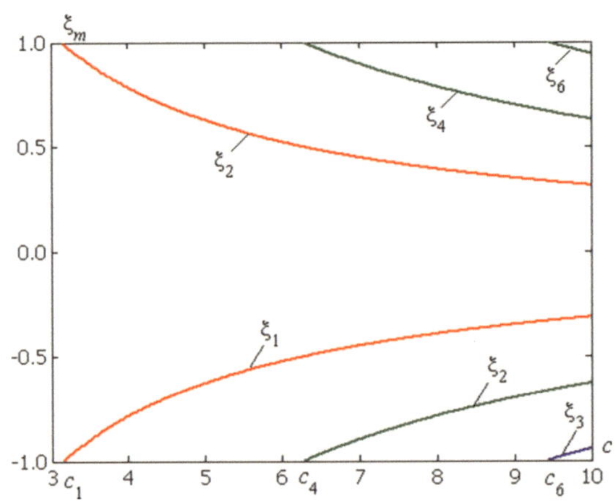

Figure 3.
Zero lines of the real solution $f_0(\xi)$.

Subsequently, the solution $\{f_{0'}, w_{0'}\}$ at point $c = c_3 \approx 6.175$ is better; by this, the curves σ_0 and $\sigma_{0'}$ cross two times.

The complex optimal solution, marked by index 1, appears at $c > c_1$. It is the symmetrical i.e. the functions $f_1(\xi)$, $w_1(x)$ (and $\varphi_1(\xi)$, $\psi_1(x)$, respectively) match in shape. The change of their phases is shown in **Figure 4** (curve 1).

Except for the real solutions (with type (a)), and the complex solution (with type (c)), the asymmetrical solutions (with type (b)) is found for the examined data. One of the determined functions (let $w(x)$) is real asymmetrical; by this, the second function, namely $f(\xi)$ has an even amplitude and an odd phase in this solutions. For the considered initial data, three mirror-symmetrical pairs of solutions exist. Let us specify the individuals of such pairs denoted by indices '2', '3', and '4', respectively (see **Figure 4**). The pair $f_2(\xi)$, $w_2(x)$ appears at the point $c = c_2$.The function $w_2(x)$ is real, and its unique zero is marked as $x_1^{(2)}$ in **Figure 2**. The function $f_2(\xi)$ has even

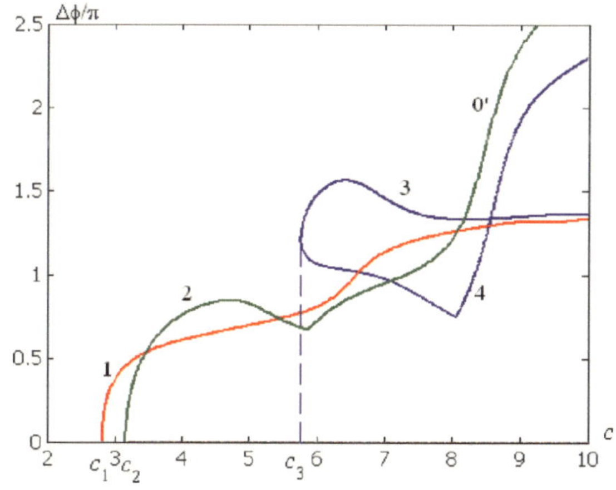

Figure 4.
Phase variations of the asymmetrical function $f_2(\xi)$.

modulus, namely $|f_2(-\xi)| = |f_2(\xi)|$, and odd phase, $|\varphi_2(-\xi)| = -|\varphi_2(\xi)|$. Its phase change $\Delta\varphi_2$ is shown as curve 2 in **Figure 4**.

The last two asymmetrical solutions marked by curves '3'and '4' are not branched off from any solutions existing before. They appear at the point $c = c_3 \approx 5.75$, and they belong to the isolated branches (**Figure 1**, curves 3 and 4). Both the functions $w_3(\xi)$ and $w_4(\xi)$ are also real. Each of the above functions has two zeros x_1, x_2 shown by the respective curves by the superscripts (3) and (4) in **Figure 2**, respectively. The changes $\Delta\varphi_3(\xi)$, $\Delta\varphi_4(\xi)$ of their odds phases are given in **Figure 4** (curves 3 and 4). One should note that for the values of c larger than $c_5 \approx 7.685$, the solution $(f_3(\xi),$ $w_3(x))$ is the best of all non-optimal solutions. It is characterized by the losses in the amount about 30% comparable to the optimal solution at the point $c = 10.0$.

4.2 Amplitude optimization

The data are given for the linear antenna (the operator A is isometric). The given amplitude DP is

$$F(\xi) = \cos(\pi\xi/2), \tag{56}$$

and the modified angular coordinate ξ is related the angular coordinate φ by $\xi = \sin\varphi/\sin\alpha$; α is angle where the DP $F(\xi)$ is non zero. Use of the modified angular coordinate ξ has some advantage, because this reduces the calculations when the adjoint operator A^* is calculated. In the case, if we use together the normalized coordinate x in the linear antenna and the modified angular coordinate ξ in a far zone, the explicit form of A^* becomes as integration over the interval $[-1, 1]$ with the complex conjugated exponent as integrand

$$u(x) = A^*g = \int_{-1}^{1} g(\xi)\exp(-icx\xi)d\xi \tag{57}$$

i.e. the operators A and A^* are pairs of two Fourier transforms having the complex conjugated exponents.

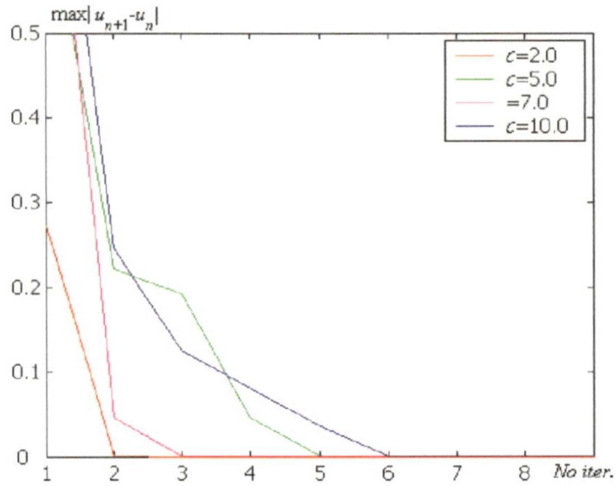

Figure 5.
The character of convergence versus number of iterations.

The iterative process (52) initiates with the approximation $u_0 \equiv 1.0$. The characteristic of convergence, namely value of $\max|u_{n+1} - u_m|$ is shown in **Figure 5**. One can conclude that only several iteration we need to attain the accuracy, which is equal to 2×10^{-14}. In spite of that the differences in the first iteration are the values 0.86, 0.69, 0.84, and 0.27 for $c = 2.0$, $c = 5.0$, $c = 7.0$, and $c = 10.0$, respectively, the high exactness 2×10^{-14} is characteristic in the sixth iteration at all values of c, which are considered.

The optimal calculated amplitudes of v for the studied parameters c are shown in **Figure 6**. The amplitude v is like constant for $c = 2.0$, and it oscillates if c grows. The above corresponds to the physical nature of such antennas that demonstrate the growth of the number of waves per length of antenna when frequency increases.

One can sure that the iterative process (47) converges quickly, and the optimal amplitudes can be easily realized for the antennas at the studied parameters of problem. The small value of the mean-square deviation σ confirms the effectiveness of the developed iterative method. This value is given for some values of c in **Figure 7**. One can see that σ does not decrease if c grows as it expected at the first view. One can conclude from

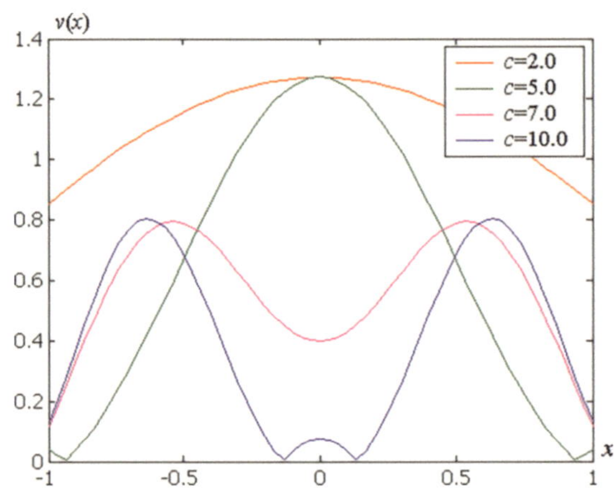

Figure 6.
The optimal amplitudes at the different values of c.

data in **Figure 7** that σ grows when c increases. The above is the result that c must correspond to the length of antenna equal to integer numbers of wave with one half of wave. Such condition is not valid for the given c. Therefore, choosing the optimal value of c is additional way to get the synthesis of results more corresponding to the practice.

In **Figure 8**, the optimal created amplitude DPs $|f|$ are marked by the color curves; the given amplitude DP F is depicted by the thick black curve. The oscillations in DPs $|f|$ are missing at smaller c, and function $|f|$ oscillates at $c = 10.0$; this is characteristic for the higher frequency radiation. The amplitudes $|f|$ are given after additional improvement related to decrease the level of radiation. Such procedure improves the approximation of amplitude $|f|$ to given function F greatly, i.e. the received values of σ functional become 0.0512, 0.0132, 0.947, and 0.1604 for $c = 5.0$, $c = 7.0$, $c = 10.0$, and $c = 12.0$, respectively. The last values of σ are considerably smaller comparably to those that are given in **Figure 7**.

The numerical data testify that the quality of approximation to the function F depends not only on the parameter c; the form of function F is important as well. The results in **Figure 9** show the optimal amplitude DPs $|f|$, which corresponds to function.

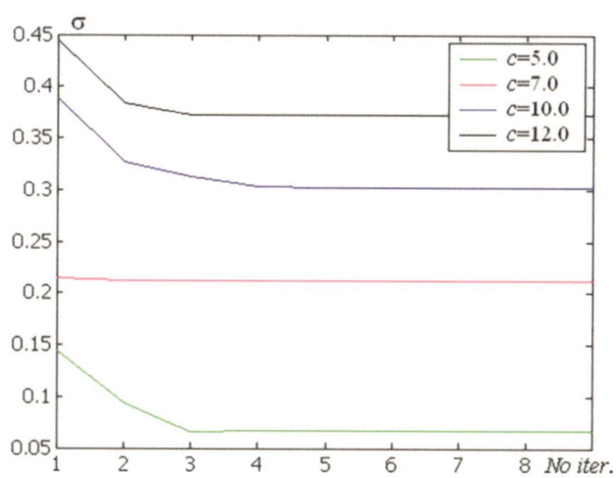

Figure 7.
The values of σ versus parameter c.

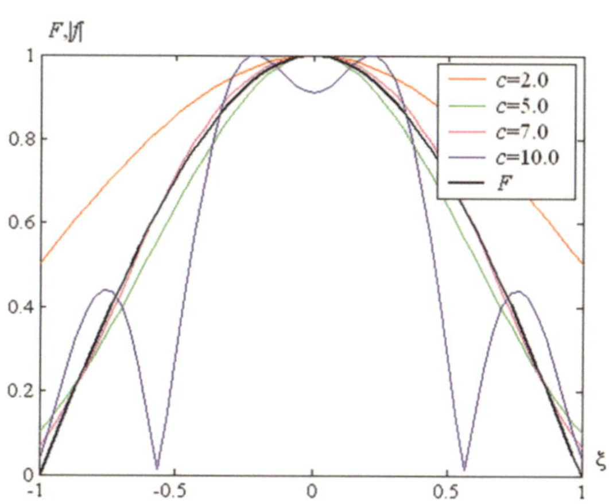

Figure 8.
The synthesized amplitude DPs at the different c.

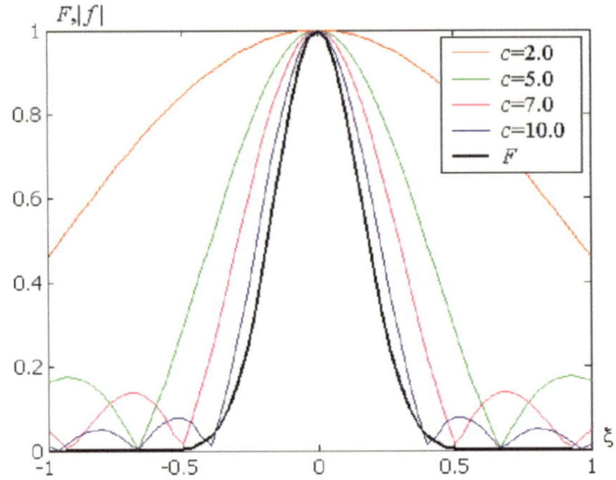

Figure 9.
The synthesized amplitude DPs for the case of narrow F.

$F(\xi) = (\cos(\pi\xi/2))^{16}$ (black thick curve). The quality of approximation to $F(\xi)$ improves if c increases. The values of (26) for this $F(\xi)$ are equal to 0.9103, 0.2014, 0.0601, and 0.0087 for the parameters $c = 2.0$, $c = 5.0$, $c = 7.0$, and $c = 10.0$, respectively.

5. Conclusions

The variational methods are applied for the formulation of the synthesis problems of antennas; and the problems the reduced to numerical solving the corresponding non-linear integral equations. The restrictions applied to the amplitude or phase of function to be optimized leads to the specific form of such equations, which in turn are solved using the method of successive approximations. The obtained results show the effectiveness of approach proposed in [17], which was used for solving the similar integral equations. The approach proposed in [5] and developed here to two types of operator A allowed to determine a set of branching solutions as well as the points of their branching. The try to consider more complex functions $F(\xi)$ demonstrates the significant computational severities that can be overcome by the modification of the proposed method of successive approximations. This will open the perspective to study of a new class of the Hammerstein non-linear integral equations and provide ability to solve the engineering synthesis problems for the different types of antennas.

Conflict of interest

The authors declare no conflict of interest.

Author details

Mykhaylo Andriychuk[1,2]

1 Pidstryhach Institute for Applied Problems of Mechanics and Mathematics, NASU, Ukraine

2 Lviv Polytechnic National University, Lviv, Ukraine

*Address all correspondence to: andr@iapmm.lviv.ua; mykhaylo.i.andriychuk@lpnu.ua

References

[1] Martins JRA, Lambe AB. Multidisciplinary design optimization: A survey of architectures. AIAA Journal. 2013;**51**(9):2049-2075. DOI: 10.2514/1.J051895

[2] Hancock BJ, Mattson CA. The smart normal constraint method for directly generating a smart Pareto set. Structural and Multidisciplinary Optimization. 2013;**48**(4):763-775. DOI: 10.1007/s00158-013-0925-6

[3] Kennedy GJ, Hicken JE. Improved constraint-aggregation methods. Computer Methods in Applied Mechanics and Engineering. 2015;**289**: 332-354. DOI: 10.1016/j.cma.2015.02.017

[4] Martins JRA, Ning A. Engineering Design Optimization. Cambridge, UK: Cambridge University Press; 2022

[5] Bulatsyk OO, Katsenelenbaum BZ, Topolyuk YP, Voitovich NN. Phase Optimization Problems. Weinheim: WILEY-VCH; 2010

[6] Andriychuk MI. Antenna Synthesis through the Characteristics of Desired Amplitude. Newcastle, UK: Cambridge Scholars Publishing; 2019

[7] Morabito AF, Rocca P. Optimal synthesis of sum and difference patterns with arbitrary sidelobes subject to common excitations constraints. IEEE Antennas and Wireless Propagation Letters. 2010;**9**:623-626. DOI: 10.1109/LAWP.2010.2053832

[8] Yang J, Yang P, Yang F, Xing Z. A hybrid approach for the synthesis of nonuniformly spaced and excited linear arrays with strict element spacing constraints. IEEE Transactions on Antennas and Propagation. 2022;**70**(7): 5521-5533. DOI: 10.1109/TAP.2022.3161525

[9] Allen OE, Wasylkiwskyj W. Antenna pattern synthesis in operational environments with electromagnetic compatibility-based constraints. IEEE Transactions on Electromagnetic Compatibility. 2004;**46**(4):668-674. DOI: 10.1109/TEMC.2004.837955

[10] Prado DR, Vaquero ÁF, Arrebola M, Pino MR, Las-Heras F. Acceleration of gradient-based algorithms for array antenna synthesis with far-field or near-field constraints. IEEE Transactions on Antennas and Propagation. 2018;**66**(10): 5239-5248. DOI: 10.1109/TAP.2018.2859915

[11] Andriychuk MI. Synthesis of plane arrays with improvement of intersystem EMC in the near and far zones. In: Proceedings of 2018 Baltic URSI Symposium (URSI); 14–17 May 2018; Poznan, Poland. New York: IEEE; 2018. pp. 178-183. DOI: 10.23919/URSI.2018.8406716

[12] Chou H-T, Wu R-Z, Akinsolu MO, Liu Y, Liu B. Radiation optimization for phased arrays of antennas incorporating the constraints of active reflection coefficients. IEEE Transactions on Antennas and Propagation. 2022;**70**(12): 11707-11717. DOI: 10.1109/TAP.2022.3209660

[13] Yang F, Yang S, Chen Y, Qu S, Hu J. Synthesis of sparse antenna arrays subject to constraint on directivity via iterative convex optimization. IEEE Antennas and Wireless Propagation Letters. 2021;**20**(8):1498-1502. DOI: 10.1109/LAWP.2021.3088492

[14] Voitovich NN, Andriychuk MI. Transformation of field in regular waveguide via phase correctors. In: Proc. of XIIth International Seminar/Workshop on Direct and Inverse

Problems of Electromagnetic and Acoustic Wave Theory (DIPED-2007); 17-21 September 2007; Lviv, Ukraine. New York: IEEE; 2007. pp. 63-66

[15] Lei S, Chen B, Lin Z, Yang W, Tian J, Hu H. Sidelobe-level minimization with power gain constraint via a wide-beam antenna array. IEEE Antennas and Wireless Propagation Letters. 2023; **22**(2):422-426. DOI: 10.1109/LAWP.2022.3214577

[16] Lin Z, Hu H, Lei S, Li R, Tian J, Chen B. Low-sidelobe shaped-beam pattern synthesis with amplitude constraints. IEEE Transactions on Antennas and Propagation. 2022;**70**(4): 2717-2731. DOI: 10.1109/TAP.2021.3125319

[17] Andriychuk MI, Voitovich NN, Savenko PA, Tkachuk VP. Synthesis of antennas according to amplitude radiation pattern. In: Numerical Methods and Algorithms. Kyiv: Naukova Dumka Publ; 1993

[18] Veinberg MM, Trenogin VA. Theory of Branching of Solutions of Non-linear Equations. Leyden: Nordhoff International Publishing; 1974

[19] Deufhard P. Newton method for nonlinear problems. Affine invariance and adaptive algorithm. In: Series of Computational Mathematics. Vol. 35. Berlin: Springer; 2004

[20] Stutzman WL, Thiele GA. Antenna Theory and Design. 3rd ed. US: John Wiley & Sons; 2012

[21] Volakis J. Antenna Engineering Handbook. 5th ed. New York: McGraw Hill; 2018

[22] Andriychuk MI, Ramm AG. Numerical solution of many-body wave scattering problem for small particles and creating materials with desired refraction coefficient. In: Awrejcewich J, editor. Numerical Simulations of Physical and Engineering Processes. Rieka, Croatia: London, UKInTech; 2011. pp. 3-28. DOI: 10.5772/24495

[23] Andriychuk MI, Voitovich NN. Antenna synthesis according to power radiation pattern with condition of norm equality. In: 2013 XVIIIth International Seminar/Workshop on Direct and Inverse Problems of Electromagnetic and Acoustic Wave Theory (DIPED); 23–26 September 2013; Lviv. New York: IEEE; 2013. pp. 137-140

[24] Andriychuk MI, Voitovich NN. Synthesis of a closed planar antenna with given amplitude pattern. Radio Engineering and Electronic Physics. 1985;**30**(5):35-40

Chapter 5

Ant Colony Optimization-Based Models for Agriculture Price Forecasting: Innovations, Case Studies, and Future Prospects

Kapil Choudhary

Abstract

Agricultural price forecasting is a critical component of modern farming, enabling stakeholders to make informed decisions. In recent years, computational intelligence techniques, such as ant colony optimization (ACO), have emerged as promising tools for enhancing the accuracy and robustness of price forecasting models. This chapter explores the innovative use of ACO-based models in agriculture price forecasting, backed by insightful case studies that demonstrate their practical applicability. The synergy between ACO and agriculture price forecasting is examined, with a focus on the unique challenges and opportunities presented by this domain. By dissecting the intricacies of these applications, we gain valuable insights into the practical implementation of ACO for agriculture price forecasting. Looking ahead, we discuss the future prospects of ACO in this field. We identify emerging trends, potential areas for improvement, and avenues for further research. The chapter concludes with a call to action for researchers, practitioners, and policymakers to collaborate in harnessing the full potential of ACO-based models, ultimately advancing the reliability and effectiveness of agriculture price forecasting. In summary, this chapter serves as a comprehensive exploration of the intersection between ant colony optimization and agriculture price forecasting. It bridges the gap between theoretical concepts and real-world applications, providing a roadmap for future advancements in this crucial domain.

Keywords: ant colony optimization, agricultural price forecasting, comprehensive, intelligence techniques, modern farming

1. Introduction

The agriculture industry plays a critical role in feeding the world's population and sustaining economic growth. Accurate price forecasting of agricultural commodities is of paramount importance for farmers, traders, policymakers, and consumers alike [1]. The ability to predict future prices enables stakeholders to make informed decisions regarding planting, harvesting, procurement, and investment, thereby mitigating risks and maximizing profits. Traditional methods of agriculture price forecasting, such as time series analysis and econometric models, have been widely used. However, they often face challenges in capturing the complex and dynamic patterns inherent in agricultural markets, which are influenced by numerous factors, including weather conditions, market demand, and geopolitical events [2]. As a result, there is a growing need for more sophisticated and adaptable forecasting techniques to address these complexities effectively. Ant Colony Optimization (ACO) [3] offers a promising avenue for tackling the challenges of agriculture price forecasting. Marco Dorigo and colleagues introduced the first ACO algorithms in the early 1990's [4–6]. The development of these algorithms was inspired by the foraging behavior of ants, ACO is a metaheuristic optimization algorithm that mimics the collective intelligence and cooperation observed in ant colonies. Originally developed to solve combinatorial optimization problems, ACO has been adapted and extended to address various real-world applications, including time series forecasting.

The core principle of ACO involves the use of artificial ants that construct solutions by iteratively exploring the solution space while laying pheromone trails on the traversed paths [7]. These pheromone trails, representing the quality of the solutions, guide other ants' choices, leading to an iterative process that converges towards promising solutions. Through this collective intelligence, ACO has demonstrated remarkable capabilities in finding optimal or near-optimal solutions in complex and dynamic problem domains. In recent years, researchers and practitioners have shown a keen interest in utilizing ACO-based models for agriculture price forecasting. This book chapter aims to provide a comprehensive overview of the innovations, case studies, and future prospects of ACO-based models in this context.

The structure of this chapter will encompass the following key aspects:

1. Fundamental principles of agriculture price forecasting, emphasizing the need for robust and accurate predictive techniques.

2. An explanation of Ant Colony Optimization, detailing its origins, principles, and adaptations for solving optimization problems.

3. The development and refinement of ACO-based models specifically tailored for agriculture price forecasting, focusing on their advantages and unique features.

4. Real-world case studies showcasing successful applications of ACO-based models in diverse agricultural commodities and regions.

5. Integration possibilities of ACO with machine learning techniques for enhanced accuracy and improved feature selection.

6. Addressing challenges related to seasonal and climate variability in agriculture price forecasting using ACO-based approaches.

7. Future prospects and potential areas of improvement for ACO-based models, fostering innovation and collaboration within the field.

By exploring these aspects, this chapter aims to shed light on the potential of ACO-based models to revolutionize agriculture price forecasting. It also emphasizes the significance of accurate price predictions in enhancing decision-making processes and ultimately contributing to the sustainable growth and stability of the agriculture sector. Through a thorough investigation of ACO's recent innovations, real-world case studies, and future prospects, this chapter seeks to inspire further research and development in the exciting intersection of Ant Colony Optimization and agriculture price forecasting.

2. Ant colony optimization (ACO) – principles and adaptations

Ant Colony Optimization (ACO) is a powerful metaheuristic inspired by ant foraging behavior [3]. Its principles, which involve collective decision-making and pheromone trail updates, have led to numerous adaptations and enhancements. ACO-based algorithms, such as MMAS, ACS, and ASrank, demonstrate improved performance in various problem domains. Moreover, the integration of ACO with machine learning and its adaptability for dynamic problems and continuous optimization further highlight its versatility and potential for solving complex real-world optimization challenges. As research in ACO continues, it is likely to unveil even more innovative adaptations and applications in the future.

2.1 ACO principles

1. **Inspiration from Ant Foraging**: ACO is inspired by the foraging behavior of real ants in their search for food as shown in **Figure 1**. Ants communicate through pheromone trails, which they deposit while exploring the environment. These trails guide other ants towards promising food sources.

2. **Optimization as a Collective Process**: ACO treats optimization as a collective decision-making process. Artificial ants construct solutions iteratively, simulating real ants' exploration and exploitation balance.

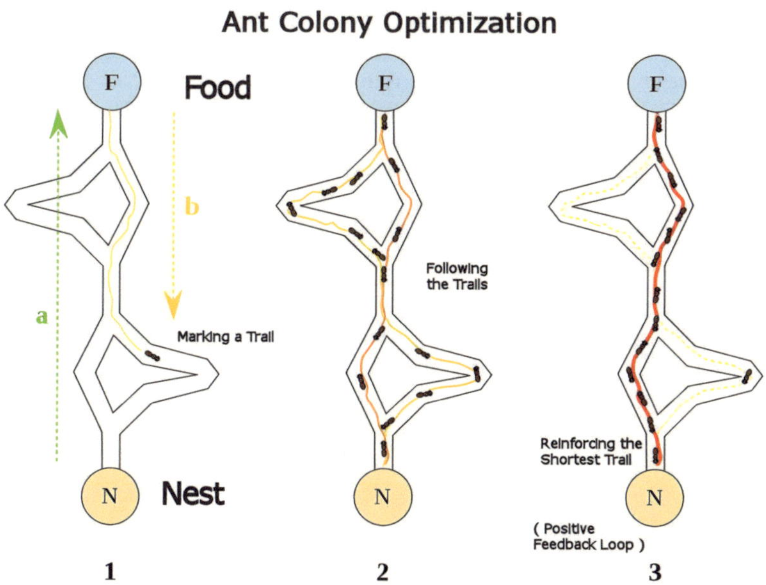

Figure 1.
Ant colony optimization.

3. **Pheromone Trail Update**: During the construction of solutions, ants deposit pheromone trails on the edges they traverse. The amount of pheromone deposited is proportional to the solution's quality. Evaporation mechanisms control the decay of pheromone trails, preventing premature convergence.

4. **Positive Feedback and Short-Term Memory**: Ants are influenced by positive feedback, preferring paths with stronger pheromone trails. Additionally, they exhibit short-term memory, making them more likely to choose paths they have recently traveled.

2.2 ACO adaptations

1. **MAX-MIN Ant System (MMAS)**: An adaptation of ACO that introduces a max-min strategy for pheromone updates. Only the best ants are allowed to deposit pheromones, preventing early convergence and promoting exploration.

2. **Ant Colony System (ACS)**: ACS enhances ACO by incorporating local search and using both deterministic and probabilistic transition rules. This improves the exploitation of promising paths.

3. **Rank-Based Ant System (ASrank)**: ASrank introduces ranking mechanisms for edge selection, considering the pheromone values of edges. This promotes diversity in the solution space and avoids premature convergence.

4. **ACO and Machine Learning Integration**: ACO has been combined with machine learning techniques to enhance its performance. It guides feature selection, parameter tuning, and model optimization.

5. **Adapting ACO for Dynamic Problems**: ACO can be adapted to handle dynamic optimization problems where the environment changes over time. Methods like Ant Colony System with Feedback (ACO-F) and Ant Colony Optimization with Random Restart (ACOR) address dynamic scenarios effectively.

6. **Hybrid ACO Models**: Hybrid ACO models combine ACO with other metaheuristics like Genetic Algorithms and Particle Swarm Optimization. These hybrid approaches leverage the strengths of different algorithms for improved optimization performance.

7. **Continuous Optimization**: Originally designed for combinatorial problems, ACO has been extended for continuous optimization tasks through discretization or the use of probability density functions.

3. ACO-based agriculture price forecasting models

ACO-Based agriculture price forecasting models leverage the principles of ant colony optimization (ACO) to predict future prices of agricultural commodities [8, 9]. These models are designed to tackle the complexities and dynamic nature of agricultural markets, enabling stakeholders to make informed decisions. Let us explore different ACO-based models used for agriculture price forecasting:

1. **Basic ACO Model**: The basic ACO model for agriculture price forecasting involves creating artificial ants that construct potential price forecast solutions. These ants explore the solution space and deposit pheromone trails on the traversed paths. The amount of pheromone deposited is proportional to the quality of the solution. Over time, ants are guided by the pheromone trails to converge towards promising price forecasts. This model provides a foundation for more advanced adaptations.

2. **MAX-MIN Ant System (MMAS)**: MMAS is an improved version of the basic ACO model. It introduces a max-min strategy for pheromone updates, where only the best ants are allowed to deposit pheromones. This strategy prevents premature convergence and promotes exploration of the solution space. MMAS enhances the algorithm's efficiency and convergence speed, making it more robust for agriculture price forecasting.

3. **Ant Colony System (ACS):** ACS is another adaptation of the basic ACO model. It combines probabilistic and deterministic transition rules, allowing ants to choose paths based on both pheromone trails and heuristic information. The inclusion of local search mechanisms improves the exploitation of promising solutions and enhances the algorithm's performance in complex price forecasting scenarios.

4. **Rank-Based Ant System (ASrank)**: ASrank is a variant that introduces ranking mechanisms for edge selection. The pheromone values of edges are ranked, and ants prefer paths with higher ranks. This approach promotes diversity in the solution space and helps avoid premature convergence. ASrank has shown promise in agriculture price forecasting, particularly in scenarios with complex price patterns.

5. **Hybrid ACO Models**: Hybrid ACO models combine ACO with other meta-heuristic algorithms or machine learning techniques. These models leverage the strengths of different algorithms to improve price forecasting accuracy. For instance, combining ACO with Genetic Algorithms or Particle Swarm Optimization can lead to enhanced performance and better exploration-exploitation balance.

6. **Continuous ACO Models**: Originally designed for combinatorial problems, ACO has been extended for continuous optimization tasks in agriculture price forecasting. Continuous ACO models use techniques like discretization or probability density functions to adapt ACO for continuous price prediction.

7. **Dynamic ACO Models**: ACO can be adapted to handle dynamic optimization problems where the environment changes over time. Methods like Ant Colony System with Feedback (ACO-F) and Ant Colony Optimization with Random Restart (ACOR) address dynamic scenarios effectively. These models are suitable for handling seasonality and climate variability in agriculture price forecasting.

Each of these models has its strengths and weaknesses, and their applicability depends on the specific characteristics of the agriculture price forecasting problem at hand. The development and integration of these ACO-based models continue to be an active area of research and innovation to achieve more accurate and reliable price predictions in the agricultural sector.

4. ACO for feature selection in agriculture price forecasting

Price forecasting in agriculture is a critical aspect of decision-making for farmers, traders, and policymakers. Accurate predictions can aid in better resource allocation, risk management, and overall sustainability of the agricultural sector. However, traditional price forecasting methods often suffer from the inclusion of irrelevant or redundant features, leading to suboptimal predictive models. ACO is a bio-inspired optimization technique based on the foraging behavior of ants. It can effectively address the feature selection challenge by intelligently selecting the most relevant features, enhancing the accuracy and efficiency of predictive models. Feature selection is crucial in predictive modeling as it helps identify the most influential attributes while discarding irrelevant ones. Conventional methods often rely on statistical measures or domain knowledge, but they may not efficiently handle high-dimensional and noisy agricultural datasets.

ACO simulates the behavior of ants when searching for the shortest path between their nest and a food source. The ants communicate through pheromone trails, reinforcing the paths that lead to food. This concept of positive feedback forms the basis of ACO algorithms. In the context of agriculture price forecasting, the feature selection problem can be formulated as an ACO task. The process involves encoding the feature subset space as a graph, where ants construct solutions by selecting a subset of features. The ants' pheromone trail represents the quality of a particular feature subset, guiding the search towards promising solutions. To apply ACO for feature selection in agriculture price forecasting, data preprocessing and preparation are essential. The ACO algorithm is integrated with machine learning models, and a case study demonstrates its efficacy in commodity price forecasting. The results showcase improved predictive accuracy and reduced computational complexity compared to traditional feature selection methods.

The performance of ACO-selected features is evaluated using standard metrics such as accuracy, precision, and recall. Additionally, the interpretability of selected features aids in understanding the driving factors behind price fluctuations, providing valuable insights for decision-makers. ACO-based feature selection offers advantages like improved model performance, reduced overfitting, and efficient handling of high-dimensional data. However, challenges such as parameter tuning and computational complexity need to be addressed for broader adoption. The application of ACO for feature selection in agriculture price forecasting shows promising potential. By selecting the most relevant features, ACO enhances the accuracy of predictive models, empowering stakeholders to make informed decisions. As research in optimization techniques continues, ACO's integration will likely play a pivotal role in advancing agriculture's productivity and sustainability in an increasingly complex market.

5. Case studies: ACO in agriculture price forecasting

Ant colony optimization (ACO) has emerged as a powerful tool in feature selection for agriculture price forecasting. In this section, we present two case studies that demonstrate the effectiveness of ACO in improving the accuracy and efficiency of predictive models for agricultural commodity prices.

5.1 Case study 1: corn price forecasting

Corn is one of the most important commodities in the agriculture sector, and accurate price forecasting is crucial for farmers and traders to optimize their operations. In this case study, historical data on corn prices, weather patterns, and other relevant factors were collected for analysis.

ACO-based Feature Selection: Using ACO, we formulated the feature selection problem by representing the feature subset space as a graph. The goal was to find the most relevant features that would contribute significantly to accurate price predictions. Ants in the ACO algorithm constructed solutions by selecting subsets of features, and their pheromone trails guided the search towards promising feature combinations.

Results: The ACO-based feature selection process effectively identified a subset of key features, including historical price trends, weather data, and economic indicators. These features were essential in capturing seasonal patterns and external factors influencing corn prices. The selected features were then used to train machine learning models, resulting in significantly improved forecasting accuracy compared to conventional methods.

Benefits: ACO-based feature selection not only enhanced predictive accuracy but also reduced computational complexity. By focusing on essential features, the model's interpretability was improved, allowing stakeholders to better understand the driving factors behind corn price fluctuations.

5.2 Case study 2: coffee price forecasting

Coffee is a highly traded commodity, and price fluctuations can have significant economic impacts on coffee-producing regions. In this case study, historical data on coffee prices, export volumes, climate variables, and global demand trends were collected and analyzed.

ACO-based Feature Selection: Similar to the first case study, ACO was applied for feature selection in coffee price forecasting. The ACO algorithm intelligently selected relevant features that captured the intricate relationships between price movements and external factors.

Results: The ACO-selected features included variables related to weather patterns, international coffee demand, and economic indicators of coffee-producing countries. By incorporating these essential features into the predictive model, the forecasting accuracy improved significantly. Additionally, the reduced feature set streamlined the model's training and evaluation process.

Benefits: The ACO-driven feature selection approach yielded a more interpretable model, enabling stakeholders to make informed decisions based on the identified driving factors behind coffee price fluctuations. Moreover, the streamlined model reduced computational resources and facilitated real-time forecasting, which is essential for time-sensitive decision-making in the coffee market.

These case studies demonstrate the effectiveness of Ant Colony Optimization in agriculture price forecasting. ACO-based feature selection enhances predictive accuracy, reduces overfitting, and improves model interpretability. By focusing on the most relevant features, ACO empowers stakeholders in the agriculture sector to make data-driven decisions, optimize resource allocation, and navigate market uncertainties successfully. As ACO continues to evolve alongside other optimization techniques, its application in agriculture price forecasting is expected to have a profound and lasting impact on the industry's productivity and sustainability.

6. Perspectives and future prospects on ACO-based models for agriculture price forecasting

Ant Colony Optimization (ACO) has shown promising results in feature selection for agriculture price forecasting, but there are still various perspectives and future prospects to explore for further advancements in this domain. In this section, we discuss potential directions and opportunities for ACO-based models in agriculture price forecasting.

1. **Enhanced Feature Selection Strategies:** Future research should focus on refining ACO's feature selection strategies to handle more complex and diverse agricultural datasets. Exploration of hybrid approaches, combining ACO with other optimization techniques, may lead to improved feature subset selection and more accurate predictions.

2. **Handling Big Data Challenges:** As the volume of agricultural data grows, ACO-based models must be adapted to handle big data challenges efficiently. Parallelization and distributed computing methods can be explored to scale ACO for large-scale agriculture price forecasting tasks.

3. **Addressing Non-Linear Relationships:** ACO currently operates based on pheromone trails, which may struggle to capture non-linear relationships between features. Investigating how to incorporate non-linearity into ACO or using ACO as a pre-processing step for non-linear models could be fruitful avenues of research.

4. **Incorporating Domain Knowledge:** Integrating domain knowledge into the ACO-based feature selection process can improve the selection of relevant features. Combining data-driven approaches with expert insights may lead to more accurate and actionable predictions for specific agricultural commodities.

5. **Expanding to Multi-Commodity Forecasting:** ACO's adaptability and efficiency make it well-suited for multi-commodity forecasting in agriculture. Future research could explore extending ACO-based models to simultaneously forecast prices for multiple commodities, considering cross-commodity dependencies and market interactions.

6. **Handling Imbalanced Data:** In agriculture price forecasting, data imbalances often occur due to fluctuating market conditions. ACO-based models should be adapted to handle imbalanced datasets, ensuring robust and accurate predictions for all scenarios.

7. **Real-time Forecasting and Decision Support:** Developing ACO-based models that can provide real-time price forecasts will be invaluable for stakeholders in agriculture. Real-time decision support can help farmers, traders, and policymakers respond promptly to market changes and make informed choices.

8. **Exploring Other Applications of ACO:** Beyond feature selection, ACO holds potential in other areas of agriculture, such as crop yield prediction, pest control optimization, and resource allocation. Exploring ACO's utility in these applications can lead to holistic improvements in agricultural practices.

9. **Benchmarking and Comparative Studies:** To assess the full potential of ACO in agriculture price forecasting, comparative studies with other state-of-the-art feature selection methods and optimization techniques are essential. Benchmarking will provide a comprehensive understanding of ACO's strengths and weaknesses in various scenarios.

10. **Adoption and Implementation**: Promoting the adoption of ACO-based models in real-world agricultural settings requires bridging the gap between research and practice. Collaboration between researchers, industry experts, and policymakers can facilitate successful implementation and ensure the models address practical challenges effectively.

ACO-based models for agriculture price forecasting present an exciting avenue for advancing the accuracy and efficiency of predictive analytics in the agricultural sector. By addressing the perspectives and exploring the future prospects discussed above, researchers can unlock the full potential of ACO in optimizing resource allocation, supporting decision-making, and ultimately contributing to the sustainable growth of agriculture in an ever-changing global market.

7. Conclusions

Ant colony optimization (ACO) has evolved significantly since its inception, leading to the development of various recent variants that have expanded its applicability across different domains. This article explored the recent variants of ACO, its diverse applications, and the promising perspectives it offers for future research and implementation. The utilization of ACO -based models for agriculture price forecasting emerges as a dynamic and effective approach, as demonstrated through diverse case studies. These models exhibit adaptability across various agricultural contexts, underscoring their versatility and robustness. With promising perspectives and ongoing research, ACO is poised to make significant contributions to solving real-world challenges and advancing optimization technology in the years to come. Looking ahead, the future prospects of this methodology are promising. Enhancements could involve the integration of additional data streams such as climatic patterns, market demand shifts, and geopolitical factors, thereby refining predictions. The potential synergies of amalgamating ACO with machine learning methods could offer an avenue for even more robust forecasting tools, will lead to more informed decision-making, improved industry efficiency, and enhanced global food security.

Acknowledgements

The author is grateful to University Grants Commission (UGC) for offering the financial assistance and also to PG School, ICAR-Indian Agricultural Research Institute, New Delhi for providing the requisite facilities to carry out this study.

Conflict of interest

The authors declare no conflict of interest.

Author details

Kapil Choudhary
Agriculture University, Jodhpur, Rajasthan, India

*Address all correspondence to: kapiliasri@gmail.com

References

[1] Wang J, Wang Z, Li X, Zhou H. Artificial bee colony-based combination approach to forecasting agricultural commodity prices. International Journal of Forecasting. 2022;**38**:21-34. DOI: 10.1016/j.ijforecast.2019.08.006

[2] Xiong T, Li C, Bao Y. Seasonal forecasting of agricultural commodity Price using a hybrid STL and ELM method: Evidence from the vegetable market in China. Neurocomputing. 2018;**275**:2831-2844. DOI: 10.1016/j.neucom.2017.11.053

[3] Dorigo M, Stutzle T. Ant Colony Optimization. Cambridge: MIT Press; 2004

[4] Dorigo M. Optimization, Learning and Natural Algorithms. Department of Electronics, Polytechnic University of Milan; 1992

[5] Dorigo M, Maniezzo V, Colorni A. Positive Feedback as a Search Strategy. Department of Electronics, Polytechnic University of Milan; 1992

[6] Dorigo M, Maniezzo V, Colorni A. Ant system: Optimization by a colony of cooperating agents. IEEE Transaction on System Management Cybernetics Part B. 1996;**26**(1):29-41

[7] Blum C. Ant colony optimization: Introduction and recent trends. Physics of Life Reviews. 2005;**2**(4): 353-373

[8] Ali M, Deo C, Xiang Y, Prasad R, Li J, Farooque A, et al. Coupled online sequential extreme learning machine model with ant colony optimization algorithm for wheat yield prediction. Scientific Reports. 2022;**12**(1):5488

[9] Huang B, Zo Q. Application of improved ant colony optimization in agricultural price forecasting. Journal of Applied Mathematics and Physics. 2016;**4**:1115-1124

Chapter 6

Application of Ant Colony Optimization in Water Resource Management

Ramadan Abdelaziz

Abstract

The use of optimization techniques in water resource management has received increasing attention in the past few decades. The main objective of optimization in water resource management is to maximize the use of the available resources while minimizing the negative impacts on the environment. Two main approaches are widely used in optimization, including global and local optimization. The main aim of the chapter is to explore the application of the Ant Colony Optimization (ACO) algorithm in water resource management. It is a flexible and powerful tool that can handle various types of hydrological models. The chapter will use different performance measures and case studies to comprehensively compare these techniques in water resource management. The results will be valuable for practitioners in choosing the appropriate optimization technique for their specific application.

Keywords: ACO, MODFLOW, MT3DMS, SWAT, HYDRUS

1. Introduction

1.1 Background on water resource management

Water is a valuable and limited resource that requires careful management to ensure its sustainable use [1]. Water resource management (WRM) involves modeling and optimization of surface water and groundwater systems to maximize water availability [2], while also minimizing conflicts between users and preserving the ecological balance [3]. The optimization process is critical in WRM because it allows decision-makers to make informed decisions based on accurate predictions and analyzes [3].

Water resource modeling is a complex process that involves the use of mathematical models to simulate the behavior of the water system [4]. These models require the estimation of parameters, such as hydraulic conductivity, recharge rate, and storage capacity [5]. The calibration of these models is essential to ensure the accuracy of the model outputs [6]. The calibration process involves adjusting the model parameters to minimize the differences between the observed and simulated data [7]. The optimization of these parameters is an important step in the model calibration process [8].

2. Optimization in water resource management

2.1 Definition of optimization

Optimization is the process of finding the best solution among a set of feasible solutions to a problem [9]. In the context of water resource management, optimization refers to the process of finding the best allocation of water resources to meet the conflicting demands of different stakeholders while considering the physical and environmental constraints of the system.

2.2 Importance of optimization in water resource management

Optimization is a crucial process in water resource management, and the choice of optimization method should be based on the complexity of the problem and the resources available [10]. Both global and local optimization methods have their advantages and disadvantages, and it is important to carefully consider these factors when choosing an optimization method for water resource management problems.

2.3 Types of optimization techniques

There are two main types of optimization methods: global optimization and local optimization. Global optimization methods aim to find the global optimum solution, which is the best possible solution that can be obtained within the search space [11]. These methods typically involve exploring the entire search space and can be computationally expensive [11, 12]. However, they are well suited for finding the global optimal solution to complex water resource management problems.

On the other hand, local optimization methods focus on finding the best solution in a specific region of the search space [13]. These methods are typically faster than global optimization methods and can be used to refine the solution found by global optimization methods. Local optimization methods are particularly useful when the search space is large, and the objective function is complex.

The choice between global and local optimization methods depends on the nature of the problem and the computational resources available. For instance, when dealing with a large water resource management system with many stakeholders and complex physical and environmental constraints, global optimization methods may be necessary to ensure that the solution found is the best possible one. However, when dealing with a smaller system, local optimization methods may be more appropriate since they can find a good solution more quickly.

3. Global optimization vs. local optimization

Global optimization and local optimization are two distinct approaches to solving optimization problems in water resource management.

3.1 Definition of global optimization

Global optimization is the process of finding the best solution within the entire search space, regardless of the starting point. It involves searching the entire solution space to identify the global optimum, which is the best solution possible. In water resource management, global optimization is used to find the best solution to a problem that may have multiple optimal solutions or where the objective function is complex.

3.2 Definition of local optimization

Local optimization, on the other hand, involves searching for the best solution in the vicinity of a given starting point. It focuses on finding the local optimum, which is the best solution that can be found within a small region of the solution space. Local optimization is used in situations where the objective function is relatively simple and the search space is small.

3.3 Differences between global and local optimization

One of the main differences between global and local optimization is the scope of the search. Global optimization involves searching the entire solution space, while local optimization only searches a small region of the solution space. Another key difference is the starting point. Global optimization does not require a starting point, whereas local optimization requires a starting point from which to begin the search.

3.4 Advantages and disadvantages of each approach

Global optimization has several advantages, including the ability to find the best possible solution, even in complex problems with multiple optimal solutions. It also has the advantage of not being affected by the starting point, making it more robust and reliable. However, global optimization can be computationally expensive and time-consuming, particularly for complex problems.

Local optimization, on the other hand, is faster and more efficient than global optimization, making it suitable for problems with simple objective functions and small search spaces. It also requires less computational resources than global optimization. However, local optimization can be affected by the starting point and may not find the global optimum.

The choice between global and local optimization in water resource management depends on the complexity of the problem, the size of the search space, and the resources available. Global optimization is suitable for complex problems with multiple optimal solutions, while local optimization is suitable for simpler problems with smaller search spaces. Both approaches have their advantages and disadvantages, and the choice between them should be based on the specific requirements of the problem at hand.

4. Applications of global optimization in water resource management

4.1 Global optimization techniques

Global optimization techniques are widely used in water resource management to find the optimal parameter set for models that involve a high degree of nonlinearity. These techniques seek to minimize the objective function by searching for a global minimum in the parameter space.

4.1.1 Ant colony optimization (ACO)

Ant colony optimization (ACO) is a metaheuristic optimization technique based on the foraging behavior of ants in nature [14]. In recent years, ACO has gained increasing popularity in solving optimization problems in various fields, including hydrology, particularly in groundwater flow and solute transport [15]. ACO has been used to solve inverse problems, which have wide applications in groundwater management, contaminant remediation, and environmental monitoring [16].

The main idea behind ACO is to simulate the behavior of ants that deposit pheromone trails as they search for food [17]. In ACO, the optimization problem is modeled as a graph, where the nodes represent the possible solutions, and the edges represent the quality of the solutions [18]. The ants follow a probabilistic decision-making process based on the pheromone trail [19] and the quality of the solutions represented by the edges [20]. The more frequently a solution is visited, the higher the pheromone concentration on the corresponding edge, and the higher the probability of choosing that solution [21].

4.1.2 Ant colony optimization in inverse modeling

The ACO algorithm is used to find the global minimum of the objective function in inverse modeling problems [22]. In the context of groundwater flow and solute transport, the objective function is typically a measure of the difference between observed and simulated data [23]. ACO is well suited to these problems because it can efficiently search a large parameter space and avoid being trapped in local optima [24, 25].

4.1.3 Application of ACO to water resource management problems

It was noted that the ACO approach has only been utilized since 2001 in the field of water resources [26]. ACO algorithms have been shown to be particularly effective for a range of optimization problems, including complex nonlinear systems [27]. Some studies have utilized ACO algorithms in the field of groundwater management and remediation [28]. For example, Li et al. [29] applied an ACO algorithm for the optimization of spatial sampling in long-term groundwater monitoring networks by reducing the number of monitoring wells while maintaining an adequate representation of groundwater conditions. Wang et al. [30] used ACO and numerical modeling to optimize remediation strategies and determine optimal locations for groundwater remedial systems.

Water distribution systems and reservoir optimization are the primary areas where ACO has been predominantly utilized. For example, Jalali et al. [31] and Kumar and Reddy [32] used the ACO algorithm to optimize reservoir operation. Jalali et al. [31] proposed an improved ACO algorithm that uses a new heuristic function and a new pheromone updating rule to improve the convergence speed and the quality of the solution. The proposed improved ACO algorithm is tested on a case study of the Hirakud reservoir in India, and the results show that it outperforms the traditional ACO algorithm and other optimization algorithms in terms of solution quality and computational efficiency. Kumar and Reddy [32] ACO technique to derive operating policies for a multi-purpose reservoir system.

Many studies utilized ACO in Water Distribution Systems. Simpson et al. [33] proposed methodology can be used to improve the performance of ACO in the optimal design of water distribution systems by selecting the parameters of ACO, including the number of ants, the heuristic information, and the pheromone evaporation rate. The proposed method tested on a case study of a water distribution system in Singapore. The results showed that the selected parameters affected the performance of the ACO algorithm. After that, Maier et al. [16] proposed modified ACO algorithm can be used to improve the performance of ACO in the design of water distribution systems. The modified ACO algorithm uses a new pheromone, a new heuristic function and updating rule to enhance the quality of the solution and the convergence speed. The proposed algorithm tested on the same case study of a water distribution system in Singapore [33]. The results proved that ACO is a promising approach for the design of water distribution systems and can be used to generate optimal designs that meet the required performance criteria. El-Ghandour and Elbeltagi [34] compare the performance of Five algorithms (GA, particle swarm optimization (PSO), DE, Artificial Bee Colony (ABC), and ACO) in terms of solution quality and computational efficiency using a case study of a water distribution network in Iran. The results showed that DE and ACO are promising approaches for the optimization of water distribution networks and can be used to generate optimal designs that meet the required performance criteria.

Ant colony optimization (ACO) has been applied in various hydrological problems, including groundwater flow and solute transport. For example, ACO is used to estimate groundwater model parameters, resulting in more accurate and reliable groundwater flow and transport predictions. Li et al. [35] proposed a hybrid ACO system that combines the traditional ACO algorithm with a local search algorithm. The hybrid system employs the ACO algorithm to generate a set of candidate solutions and the Simulated Annealing (SA) algorithm to refine the solutions by accepting some of the worst solutions with a certain probability. The proposed hybrid ACO system is tested on a case study of a groundwater system in China. The proposed hybrid ACO system is a promising approach for parameter estimation in groundwater hydrology and can lead to more accurate and reliable predictions of groundwater flow and transport. Irani and Nasimi [36] proposed ACO algorithm used to improve the performance of neural networks in the estimation of the permeability of a reservoir. The proposed new method is called ACA-BP, which combines the ACO algorithm with the Backpropagation (BP) algorithm. The method was tested on a case study of a reservoir in Iran. The proposed ACA-BP method is a promising approach for the estimation of the permeability of a reservoir and can be used to generate accurate and reliable predictions of the performance of neural networks in the estimation of the permeability of a reservoir. Dobre and Drobot [37] proposed ACO-Artificial Neural Network (ANN) method can be used to improve the accuracy and efficiency of the estimation of soil hydraulic parameters, which are important for the prediction of soil water movement and plant growth. A new proposed method called ACO-ANN, which combines the ACO algorithm with the Artificial Neural Network (ANN) to improve the accuracy and efficiency of the estimation process. The proposed method is tested on a case study of soil in Romania.

Ghorbani et al. [38] used ACO algorithm to solve the fuzzy optimization problem for river water quality management. The proposed model handled multiple objectives and multiple pollutants in a river system. The proposed model is tested on a case study of Sefid-Roud River in Iran and uses a fuzzy optimization model to determine the optimal waste load allocation for the pollutants in the river. The proposed model could effectively allocate pollutant loads, leading to improved river water quality. Guo et al. [39] applied ACO algorithm to synthetic and field 3D resistivity imaging examples. In both synthetic and field, the *a priori* constrained ACO method obtained better models with higher resolution and accuracy. The *a priori* constraints were based on geological knowledge of the geological feasibility, survey area, and enforce physical on the resistivity models. The constraints applied on thickness of geological layers, resistivity range, and layer sequence. The results showed that incorporating prior geological constraints into the ACO algorithm can significantly improve imaging quality and feasibility for 3D electrical resistivity imaging applications.

The ACO method developed by Jamei et al. [40] was utilized as one of the techniques to assess the salinity of multi-aquifers in Bangladesh's coastal regions. The ACO model used a new hybrid neuro-computing approach that combined the ACO algorithm with the Adaptive Neuro-Fuzzy Inference System (ANFIS) and the Slime Mould Algorithm (SMA) to estimate the spread of salinity in Bangladesh's coastal aquifers. The new hybrid neuro-computing approach demonstrates a promising approach for assessing groundwater salinity distribution and can be utilized to provide accurate and reliable projections of the salinity trend in Bangladesh's coastal groundwater. Ahmed et al. [41] developed two novel deep learning (DDDL) algorithms for forecasting river water levels: DDDL-ACO and DDDL-PSO. These approaches optimize the structure of the deep neural network using ant colony optimization (ACO) and particle swarm optimization (PSO), respectively. The ACO feature selection technique was utilized to determine the most important input variables from climatic indices, satellite data, and ground-based data. The fuzzy multi-objective model introduced by Ghorbani et al. [42] in conjunction with the ACO algorithm is an excellent decision-support tool for reducing numerous contaminants in rivers. The ACO algorithm iteratively seeks optimal pollution reduction solutions that meet the model's objectives. The study discovers that the ACO algorithm may generate a collection of Pareto optimal solutions that represent the trade-offs between the objectives. The ACO algorithm outperforms other evolutionary algorithms in terms of convergence speed, solution variety, and computational efficiency. It is capable of dealing with the uncertainties and complexities of real-world river pollution concerns. For the bivariate simulation of river flows in the subbasins of Lake Urmia, Eslamitabar et al. [43] suggest the hybrid intelligence models ANFIS-ACO and ANFIS-PSO. These correspondingly combine ant colony optimization (ACO) and particle swarm optimization (PSO) with an adaptive neuro-fuzzy inference system (ANFIS). The ANFIS model's parameters are optimized using the ACO and PSO algorithms. They look for the ideal parameter setting that reduces the discrepancy between the simulated and actual flow data. According to the study, the ANFIS-ACO model outperforms the ANFIS-PSO model in terms of accuracy and simulation error for both low and high river flows. The ANFIS-ACO model performs better than the study's individual and hybrid models as well. This shows that PSO and ACO are both capable of optimizing the parameters of the ANFIS model.

Another application of ACO in Optimal Crop and Irrigation Water Allocation, Nguyen et al. [44], offered an enhanced ant ACO formulation for the allocation of crops and water to different irrigation areas as another use of ACO. The improved ACO formulation is compared with that of other ACO algorithm variants (without and with domain knowledge) for two case studies, including one from the literature and one introduced in the paper for different water availability scenarios within an irrigation district located in Loxton, South Australia, near the River Murray. The revised ACO formulation with domain knowledge is a viable technique for optimal crop and irrigation water allocation problems, particularly those requiring more complex crop models than those utilized in the case studies.

4.1.4 Advantages of ACO

Ant colony optimization (ACO) has several advantages over other optimization methods. One of the main advantages is that it can efficiently explore a large parameter space without getting trapped in local optima. This is particularly useful in inverse modeling problems where there are often multiple local optima. Another advantage is that ACO can handle discrete and continuous parameters, which makes it useful for problems with mixed variables. Additionally, ACO has been found to be computationally efficient and scalable.

4.1.5 Limitations of ACO

Despite its advantages, ACO also has some limitations. One of the main limitations is that it requires the definition of a pheromone update rule. The pheromone update rule is critical for the algorithm's performance, but it can be difficult to determine the optimal rule for a particular problem. Additionally, ACO is prone to premature convergence, which can limit its ability to explore the parameter space.

4.1.6 Conclusion

Ant colony optimization (ACO) is a promising and effective optimization tool for inverse modeling of groundwater flow and solute transport. It has been found to be effective in finding the global minimum of the objective function in complex problems. Its ability to efficiently search large solution spaces and its adaptability to different problem types make it a valuable tool for hydrogeologists and other researchers in related fields. However, careful consideration must be given to the selection and tuning of its parameters, and researchers should be aware of its limitations and explore hybrid and modified versions of the algorithm to improve its performance in specific applications. Overall, ACO offers a powerful and efficient alternative to traditional optimization methods for solving complex inverse modeling problems. Its ability to search the parameter space efficiently and effectively makes it a suitable option for optimizing complex and computationally expensive hydrological models. However, further research is needed to fully understand its capabilities and limitations and to explore its potential applications in other areas of hydrological modeling. In addition, further research is required to address the limitations of the algorithm and optimize its performance to fully understand its capabilities and limitations in different contexts. Moreover, the effectiveness of ACO can be affected by various factors, such as the choice of parameters and the complexity of the model.

4.2 ACO algorithms' fundamental mathematical equations

These mathematical expressions aid in the understanding of the fundamental optimization principles behind techniques such as ACO algorithm. Specific formulations then differ depending on the algorithms and applications.

The basic mathematical equations for ACO algorithm is provided here:

4.2.1 ACO

Global optimization technique inspired on ant foraging behavior [45].

$$P_{i,j} = \frac{\left(\tau_{i,j}^{\alpha}\right)\left(\eta_{i,j}^{\beta}\right)}{\sum\left(\tau_{i,j}^{\alpha}\right)\left(\eta_{i,j}^{\beta}\right)} \tag{1}$$

where
$P_{i,j}$ is the probability of choosing path j from i.
$\tau_{i,j}^{\alpha}$ is the pheromone trail concentration.
$\eta_{i,j}^{\beta}$ is the heuristic value (inverse of distance).
α and β are parameters to control influences for $\tau_{i,j}^{\alpha}$ and $\eta_{i,j}^{\beta}$.

4.2.2 Local pheromone update

After an ant traverses an edge (i, j), the pheromone level on that edge is locally updated as follows:

$$\tau_{i,j} = (1-\rho)\tau_{i,j} + \rho\Delta\tau_{i,j} \tag{2}$$

where ρ is the pheromone evaporation rate, and $\Delta\tau_{i,j}$represents the amount of pheromone deposited by the ant.

4.2.3 Global pheromone update

After all ants complete their tours, the pheromone level on the edges is globally updated based on the quality of the solutions found. The update is performed as follows [46]:

$$\tau_{i,j} = (1-\rho)\tau_{i,j} + \sum\left(\Delta\tau_{i,j}^{k}\right) \tag{3}$$

where $\Delta\tau_{i,j}^{k}$ represents the amount of pheromone deposited by ant k on edge (i, j).

4.3 Recommendations

The optimization approach used should be determined by the characteristics of the water resource problem, such as model type, data, objective function, restrictions, and uncertainty. The problem complexity, initial parameter estimations, objective function, computing demand, desired optimality, trade-off analysis, and hybrid approaches should all be taken into account while selecting the ACO algorithm for managing water resources. If these parameters are carefully taken into account, the ACO algorithm can offer practical answers to challenging optimization issues in the management of water resources. Here are some suggestions based on various viewpoints:

4.3.1 Problem complexity

Ant colony optimization (ACO) may not be appropriate for issues with a very wide or dynamic search space, or for situations that demand accurate answers. ACO, on the other hand, can handle complicated problems with discrete or continuous variables, nonlinearities, numerous objectives, and restrictions.

4.3.2 Initial parameter estimates

If initial parameter estimations are far from ideal, it is more likely to find the genuine optimum. ACO can explore the search space and adapt to the problem's features, therefore, it does not need much prior information or initial parameter estimates. Nevertheless, ACO would need some tuning of its parameters, such as the number of ants and the relative importance of pheromones and heuristic data.

4.3.3 Objective function

Ant colony optimization (ACO) may collaborate with any objective function that assesses the quality of a solution. ACO could work better with objective functions that are smooth, continuous, and differentiable, though, since they can provide the ants more information to help them with their search. In order to capture various parts of the issue, ACO may potentially profit from the use of several objective functions.

4.3.4 Computational demand

Ant colony optimization (ACO) is a population-based algorithm. ACO could also be computationally costly because it takes several iterations and evaluations to converge to a satisfactory solution. ACO is a relatively computationally demanding algorithm. The computational demand can be reduced by using parallelization techniques and distributed computing. If the pheromone trails become too dominant or too weak, ACO may potentially experience stagnation or premature convergence.

4.3.5 Desired optimality

Ant colony optimization (ACO) is a metaheuristic algorithm that can discover approximations of solutions that are sufficient for use in real-world situations. ACO does not promise to identify the global or even a local optimum. ACO could also have trouble identifying other options or escape from local optima. It is not a given that ACO will identify a problem's ideal resolution. ACO, however, frequently finds effective solutions to issues that other optimization methods find challenging. ACO is not always the best algorithm. However, it frequently comes up with effective answers to issues with water resource management.

4.3.6 Trade-off analysis

Think about the trade-offs that exist between solution quality and computing efficiency. ACO is appropriate for multi-objective optimization or issues with complicated trade-offs because it strikes a solid balance between exploration and exploitation. To select the best method, weigh the trade-offs between solution quality, computational efficiency, and resource availability. When managing water resources, ACO may be used to undertake trade-off analyzes between several goals. ACO, for instance, may be used to identify a solution that maximizes water supply while minimizing the cost of water treatment. ACO can be used to balance computing demand with solution optimality. This makes it a suitable alternative for cases where finding a decent solution fast is critical. ACO can be used to trade-off multiple goals, such as lowering costs while enhancing water quality. As a result, ACO is a useful technique for water resource management problems with various objectives.

4.3.7 Hybrid methods

Ant colony optimization (ACO) can be hybridized or combined with other methods to enhance its efficacy and flexibility. To improve the solutions discovered by the ants, for instance, ACO can employ local search or other heuristics. It can also use additional metaheuristics or evolutionary algorithms to diversify the search or improve convergence. For particular water resource management issues, hybrid techniques can improve performance by combining the advantages of several algorithms. These frequently offer greater performance than relying just on ACO. As a result, it is a flexible tool that may be used to a variety of water resource management issues.

Ant colony optimization (ACO) has various disadvantages, such as convergence to the best solution can be slow, especially for big and complicated problems. Furthermore, performance is dependent on parameter adjustment, which might be problem-specific. ACO might be a fantastic starting point for finding high-quality solutions in a timely manner. Global optimization techniques can be used to refine solutions and strive for the accurate global optimum if more accurate optimization is required.

Finally, in water resource management, the decision between ACO and global optimization is determined by the specific issue being handled and the computational resources available. ACO can be a strong and adaptable strategy to tackling complicated water resource management problems, although global optimization approaches may be better suited to situations requiring the global minimum of a complex objective function. The issue complexity, computing demand, intended optimality, and trade-off analysis should all be considered while selecting an algorithm such as ACO as global optimization approaches is well suited for solving complex and nonlinear water resource problems with many objectives, constraints, and uncertainties.

Author details

Ramadan Abdelaziz
gbm Society for Building Geology and Measurement Technology Building Ground Institute, Ettlingen, Germany

*Address all correspondence to: ramawaad@gmail.com

References

[1] Water Resources Management. World Bank. Available from: https://www.worldbank.org/en/topic/waterresourcesmanagement; [Accessed: April 03, 2023]

[2] Hart S. Water Resources Management: Principles and Practice [Online]. Syrawood Publishing House; 2022. Available from: https://books.google.de/books?id=pmO1zgEACAAJ

[3] Loucks DP, van Beek E, Stedinger JR, Dijkman JPM, Villars MT. Water Resources Systems Planning and Management: An Introduction to Methods, Models and Applications [Online]. Paris: UNESCO; 2005. Available from: https://ecommons.cornell.edu/handle/1813/2804 [Accessed: April 02, 2023]

[4] Vedula S, Mujumdar PP. Water Resources Systems: Modelling Techniques and Analysis [Online]. Bangalore: Indian Institute of Science (IISc); 2014. Available from: http://civil.iisc.ac.in/~pradeep/wp-content/uploads/2020/10/Vedula-and-Mujumdar_Book_Water-Resources-Systems.pdf

[5] Sophocleous M. Groundwater recharge and sustainability in the high plains aquifer in Kansas, USA. Hydrogeology Journal. 2005;**13**:351-365

[6] Beven KJ. Rainfall-Runoff Modelling: The Primer. John Wiley & Sons; 29 Nov 2011. Available from: https://books.google.de/books?hl=en&lr=&id=eI-jjlTirlAC&oi=fnd&pg=PR7&dq=Rainfall-runoff+modelling:+the+primer&ots=9TEO0SYp29&sig=KEpz1HWOieEfW9PckSbebHZKRUM&redir_esc=y#v=onepage&q=Rainfall-runoff%20modelling%3A%20the%20primer&f=false

[7] Duan Q, Sorooshian S, Gupta V. Effective and efficient global optimization for conceptual rainfall-runoff models. Water Resources Research. 1992;**28**(4):1015-1031

[8] Moriasi DN, Arnold JG, Van Liew MW, Bingner RL, Harmel RD, Veith TL. Model evaluation guidelines for systematic quantification of accuracy in watershed simulations. Transactions of the ASABE. 2007;**50**(3):885-900

[9] Boyd SP, Vandenberghe L. Convex Optimization. Cambridge University Press; 8 Mar 2004. Available from: https://books.google.de/books?hl=en&lr=&id=mYm0bLd3fcoC&oi=fnd&pg=PR11&dq=Convex+optimization&ots=tfbWvBGBO0&sig=ZShETSc6pfmPOKb0toqf6FC2kaE&redir_esc=y#v=onepage&q=Convex%20optimization&f=false

[10] Giuliani M, Castelletti A. Assessing the value of cooperation and information exchange in large water resources systems by agent-based optimization. Water Resources Research. 2013;**49**(7):3912-3926. DOI: 10.1002/wrcr.20287

[11] Pardalos PM, Romeijn HE, editors. Handbook of Global Optimization: Volume 2. Springer Science & Business Media; 18 Apr 2013. Available from: https://citeseerx.ist.psu.edu/document?repid=rep1&type=pdf&doi=05d32628d1fc2b8afd3e540bae4b7c31b71d7079

[12] Horst R, Pardalos PM, editors. Handbook of Global Optimization. Springer Science & Business Media; 11 Dec 2013. Available from: https://books.google.de/books?hl=en&lr=&id=yBDaBwAAQBAJ&oi=fnd&pg=PA1&dq=Handbook+of+global+optimization+ability&ots=NpoMMpoAGG&sig=o7Bet_dDDR-BvaxUwp4ut0vyLII&redir_esc=y#v=onepage&q=Handbook%20of%20global%20optimization%20ability&f=false

[13] Maier HR et al. Evolutionary algorithms and other metaheuristics in water resources: Current status, research challenges and future directions. Environmental Modelling & Software. 2014;**62**:271-299

[14] Dorigo M. Positive Feedback as a Search Strategy. Technical Report. 1991. pp. 91-16

[15] Reddy MJ. Swarm intelligence and evolutionary computation for single and multiobjective optimization in water resource systems [Doctoral dissertation]. 2008. Available from: https://etd.iisc.ac.in/handle/2005/370

[16] Maier HR et al. Ant colony optimization for design of water distribution systems. Journal of Water Resources Planning and Management. 2003;**129**(3):200-209

[17] Dorigo M, Birattari M, Stutzle T. Ant colony optimization. IEEE Computational Intelligence Magazine. 2006;**1**(4):28-39

[18] Dorigo M. Optimization, learning and natural algorithms [PhD thesis]. Politecnico di Milano; 1992

[19] Dorigo M, Gambardella LM. Ant colony system: A cooperative learning approach to the traveling salesman problem. IEEE Transactions on Evolutionary Computation. 1997;**1**(1):53-66

[20] Dorigo M, Maniezzo V, Colorni A. Ant system: Optimization by a colony of cooperating agents. IEEE Transactions on Systems, Man, and Cybernetics Part B: Cybernetics. 1996;**26**(1):29-41

[21] Dorigo M, Di Caro G. Ant colony optimization: A new meta-heuristic. In: Proceedings of the 1999 Congress on Evolutionary Computation-CEC99 (Cat. No. 99TH8406). Vol. 2. IEEE; 6 Jul 1999. pp. 1470-1477. Available from: http://www.cs.yale.edu/homes/lans/readings/routing/dorigo-ants-1999.pdf

[22] Carvalho AR, de Campos Velho HF, Stephany S, Souto RP, Becceneri JC, Sandri S. Fuzzy ant colony optimization for estimating chlorophyll concentration profile in offshore sea water. Inverse Problems in Science and Engineering. 2008;**16**(6):705-715

[23] Zhou H, Gómez-Hernández JJ, Li L. Inverse methods in hydrogeology: Evolution and recent trends. Advances in Water Resources. 2014;**63**:22-37

[24] Adrian AM, Utamima A, Wang K-J. A comparative study of GA, PSO and ACO for solving construction site layout optimization. KSCE Journal of Civil Engineering. 2015;**19**:520-527

[25] Khosravanian R, Mansouri V, Wood DA, Alipour MR. A comparative study of several metaheuristic algorithms for optimizing complex 3-D well-path designs. Journal of Petroleum Exploration and Production Technologies. 2018;**8**:1487-1503

[26] Bhavya R, Elango L. Ant-inspired metaheuristic algorithms for combinatorial optimization problems in water resources management. Water. 2023;**15**(9):1712

[27] López-Ibáñez M, Prasad TD, Paechter B. Ant colony optimization for optimal control of pumps in water distribution networks. Journal of Water Resources Planning and Management. 2008;**134**(4):337-346

[28] Li Y, Hilton ABC. Optimal groundwater monitoring design using an ant colony optimization paradigm. Environmental Modelling & Software. 2007;**22**(1):110-116

[29] Li Y, Hilton ABC. Reducing spatial sampling in long-term groundwater monitoring networks using ant colony optimization. International Journal of Computational Intelligence Research. 2005;**1**(1):9-28

[30] Wang Z, Lu W, Chang Z, Wang H. Simultaneous identification of groundwater contaminant source and simulation model parameters based on an ensemble Kalman filter–adaptive step length ant colony optimization algorithm. Journal of Hydrology. 2022; **605**:127352

[31] Jalali MR, Marino M, Afshar A. Improved ant colony optimization algorithm for reservoir operation. Scientia Iranica. 2006;**13**(3):295-302. Available from: https://www.researchgate.net/publication/288701904_Improved_ant_colony_optimization_algorithm_for_reservoir_operation

[32] Kumar DN, Reddy MJ. Ant colony optimization for multi-purpose reservoir operation. Water Resources Management. 2006;**20**:879-898

[33] Simpson AR, Maier HR, Foong WK, Phang KY, Seah HY, Tan CL. Selection of parameters for ant colony optimization applied to the optimal design of water distribution systems. In: Proc. Int. Congress on Modeling and Simulation, Canberra, Australia. Dec 2001. pp. 1931-1936. Available from: https://www.researchgate.net/profile/Angus-Simpson/publication/270824003_Selection_of_parameters_for_Ant_Colony_optimisation_applied_to_the_optimal_design_of_water_distribution_systems/links/54b5fc4a0cf2318f0f9a06ef/Selection-of-parameters-for-Ant-Colony-optimisation-applied-to-the-optimal-design-of-water-distribution-systems.pdf

[34] El-Ghandour HA, Elbeltagi E. Comparison of five evolutionary algorithms for optimization of water distribution networks. Journal of Computing in Civil Engineering. 2018; **32**(1):04017066

[35] Li S, Liu Y, Yu H. Parameter estimation approach in groundwater hydrology using hybrid ant colony system. In: Computational Intelligence and Bioinformatics: International Conference on Intelligent Computing, ICIC 2006, Kunming, China, August 16-19, 2006. Proceedings, Part III 2. Berlin, Heidelberg: Springer; 2006. pp. 182-191

[36] Irani R, Nasimi R. An evolving neural network using an ant colony algorithm for a permeability estimation of the reservoir. Petroleum Science and Technology. 2012;**30**(4):375-384

[37] Dobre GR, Drobot R. Estimating soil hydraulic parameters using a metaheuristic algorithm. Mathematical Modeling in Civil Engineering. 1 Mar 2011;**1-2**:102-111. Available from: https://www.researchgate.net/profile/Radu-Drobot/publication/268323424_ESTIMATING_SOIL_HYDRAULIC_PARAMETERS_USING_A_METAHEURISTIC_ALGORITHM/links/5689510108ae051f9af762c3/ESTIMATING-SOIL-HYDRAULIC-PARAMETERS-USING-A-METAHEURISTIC-ALGORITHM.pdf

[38] Ghorbani MK, Afshar A, Hamidifar H. River water quality management using a fuzzy optimization model and the NSFWQI index. Water SA. 2021;**47**(1):45-53

[39] Guo Q et al. A priori constrained ACO method applied to three-dimensional imaging of subsurface electrical resistivity. Geotechnical and Geological Engineering. 2023;**41**(1):15-25

[40] Jamei M, Karbasi M, Malik A, Abualigah L, Islam ARMT, Yaseen ZM.

Computational assessment of groundwater salinity distribution within coastal multi-aquifers of Bangladesh. Scientific Reports. 2022;**12**(1):11165

[41] Ahmed AM et al. New double decomposition deep learning methods for river water level forecasting. Science of the Total Environment. 2022;**831**: 154722

[42] Ghorbani MK, Afshar A, Hamidifar H, Reddy MJ. A fuzzy multi-objective multiple-pollutant model for rivers using an ant colony algorithm. In: Proceedings of the Institution of Civil Engineers-Water Management. Vol. 175, No. 4. Thomas Telford Ltd. ICE Virtual Library; Aug 2022. pp. 190-205. Available from: https://www.icevirtuallibrary.com/doi/epdf/10.1680/jwama.20.00081

[43] Eslamitabar V, Ahmadi F, Sharafati A, Rezaverdinejad V. Bivariate simulation of river flow using hybrid intelligent models in sub-basins of Lake Urmia, Iran. Acta Geophysica. 2023; **71**(2):873-892

[44] Nguyen D, Dandy G, Maier H, Ascough J. Improved ant colony optimization for optimal crop and irrigation water allocation by incorporating domain knowledge. Journal of Water Resources Planning and Management. 2016;**142**(9):04016025

[45] Mosa MA, Hamouda A, Marei M. Ant colony heuristic for user-contributed comments summarization. Knowledge-Based Systems. 2017;**118**: 105-114

[46] Ning J, Zhang Q, Zhang C, Zhang B. A best-path-updating information-guided ant colony optimization algorithm. Information Sciences. 2018; **433**:142-162

Section 2

Optimization in Information Technology

Chapter 7

Fast Particle Search and Positioning Algorithms Using an Efficient Cell Registration Method

Yoshifumi Ogami and Siddique Kamran

Abstract

Cartesian Cell Registration (CCR) is an effective method to reduce computational time for particle search and positioning, for example, in the Direct Simulation Monte Carlo (DSMC) method for dilute fluid flows and multiphase flow calculations. In this chapter, an efficient FORTRAN algorithm for the CCR method is presented to further reduce both the computational time for registration and computer memory. With this algorithm, the computation time for searching the target cell where the target particle exists is reduced by as much as 44,234 times. Moreover, this algorithm was successfully applied to the DSMC simulation, resulting in a 369-fold reduction in computational time compared to the brute-force approach, that is, searching all target cells until the cell in which the target particle is present is found.

Keywords: Cartesian cell registration, direct simulation Monte Carlo, particle searching, area comparison, FORTRAN algorithm

1. Introduction

Particle search and positioning are the main time-consuming processes in the direct simulation Monte Carlo (DSMC) method [1, 2] for dilute fluid flows and the calculation of multiphase flows [3–5]. When a Cartesian (rectangular) grid system is used for these simulations, particle search and positioning are quite straightforward because the identification number (ID number) associated with the target cell in which a target particle is located can be calculated using a simple algebraic expression. However, structured and unstructured grid systems require a time-consuming process for searching a large number of target cells to position the target particles, although these grid systems can simulate flows around complex geometries with better quality of body fitting and better accuracy of object boundary calculation than the Cartesian grid system [6].

Taking advantage of the Cartesian grid system to efficiently search for target cells, Liang et al. [7] proposed a DSMC method that combines two levels of Cartesian grid systems with an unstructured triangular grid system. However, as Wang et al. [8] point out that although this method improves the calculation efficiency, multiple calculations must be performed separately at the junction of the two types of grids, which can be relatively tedious. Wang C et al. [9] divided the computational domain into rectangular regions for Cartesian grid systems by using a small number of triangular elements. The trajectory of the target particle is first tracked by searching the rectangular regions, followed by the triangular elements. This method also improves the calculation efficiency. However, the target grid cell system is limited to a triangular grid system, and searching for target grids in small areas can be time-consuming.

Wang Z et al. [8] proposed a Background Cartesian Grid Positioning (BCP) technique for two-dimensional computations. In this technique, a structured grid system is superposed on a layer of a background Cartesian grid system, and the geometric relationship between the structured and background grids is determined in advance to limit the number of target cells to be searched later. Although the results showed a significant reduction in the computational time required for particle positioning, the BCP was limited to structured grid systems as target grid systems. In addition, applying BCP to three-dimensional grids can be challenging.

To overcome these limitations, Ogami [10] introduced two methods to limit the target cells to search for the positioning of a target particle in a structured or unstructured grid system. The first is the Surrounding Cell Registration (SCR) method. In this method, the ID numbers of the target cells surrounding the central target cell are registered in advance with the central target cell to limit the cells to be searched for later. It has been demonstrated that a program using this method runs faster than other methods, as SCR has the smallest number of cells to search among the methods studied. However, SCR has two disadvantages: the ID numbers of the target cells into which the particles are introduced must be known in advance or predetermined, and it cannot be applied to particles that move through more than two cells during one-time step.

These drawbacks are overcome by the second method, called Cartesian cell registration (CCR). CCR uses a Cartesian grid system as a background two-dimensional grid system, as in the method proposed by Wang et al. [8]. However, this method can be applied in a completely different way to both structured and unstructured grids as target grids. Using SCR and CCR for a two-dimensional potential flow around a circular cylinder as an application of the proposed methods, it is shown that computational programs using both methods are almost 200 times faster than a program using the brute-force approach (searching all target cells until the target particle is found).

The ideal condition for CCR is that only one target cell is registered in each cell of the Cartesian grid system. To achieve this, the cell size of the Cartesian grid system must be as small as the smallest cell of the target grid system, in which the flow is to be simulated. However, as the number of cells of the Cartesian grid system increases (while the cell size of the Cartesian grid system decreases), the computer time for registration and the use of computer memory increase immeasurably. To solve this problem, we propose an efficient computer algorithm for registration. Using this method, simulation results of the DSMC method for dilute flows around two-dimensional circular cylinders are presented. It was found that the program for searching target cells (finding the ID number of a target cell) in which target particles exist runs 44,234 times faster, and the entire program, including the search program and DSMC simulation, runs 448 times faster than the brute-force approach.

2. CCR method and the problem of previous FORTRAN algorithm

In this section, we explain the problem of the FORTRAN program for Cartesian cell registration (CCR) method [10]. Consider a grid system around a circular cylinder as the target grid system (shown in red in **Figure 1**), which is to be registered with a Cartesian grid system (shown in black in **Figure 1**). The target grid system can be structured, unstructured, or have a polygonal structure that is not a quadrangular structure. Also, the ID placed on the target cells were either sequential or arbitrary.

Consider a square cell S_0 of width W drawn in bold black lines on the Cartesian grid system, as shown in **Figure 2**. The ID numbers of all red cells in the target grid

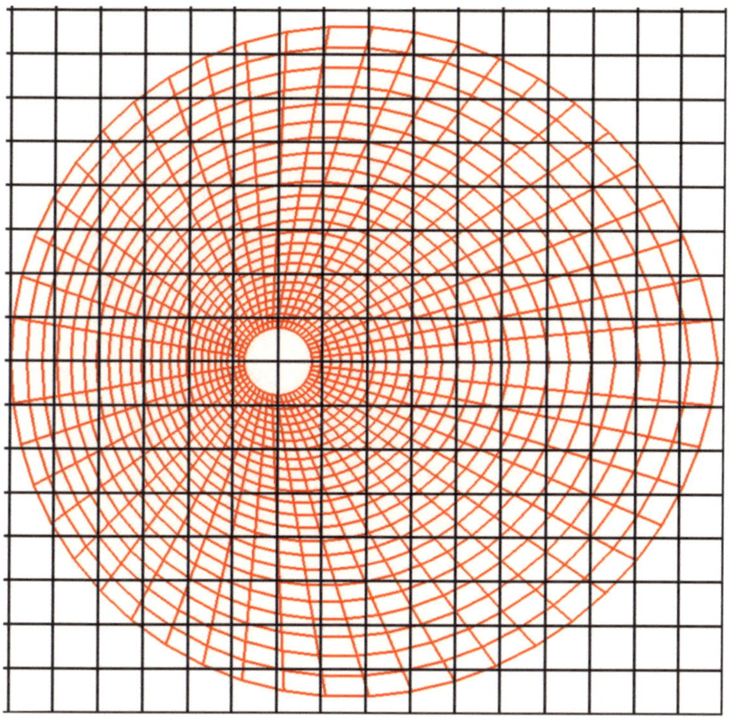

Figure 1.
Target grid cell system (red) and Cartesian grid cell system (black).

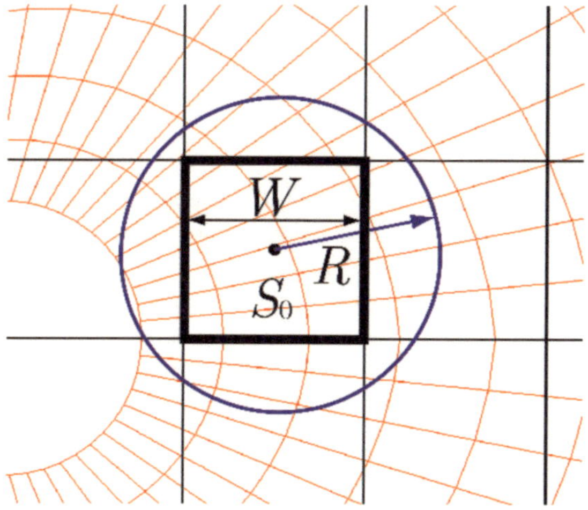

Figure 2.
Target red cells in the blue circle are registered with the black Cartesian grid cell So.

system whose centers are within a distance R from the center of S_0 are registered with the square cell S_0. This registration process was performed for all Cartesian and target red cells. Some red target cells may have been registered more than once with different Cartesian grid cells. However, it is very important to adjust parameters W and R so that no red target cells remain unregistered. The number of registered target cells may be zero or excessive, depending on the W and R parameters and the respective cell size of the target grid system.

The core components of the FORTRAN algorithm for CCR are shown in Appendix I (the author is familiar with the old FORTRAN).

Suppose the x and y coordinates of the target particle are xp and yp, respectively. The ID number of the target cell in which this particle is located can be determined using the program in Appendix II.

The above program can be used to determine the ID number of the target cell. Using this algorithm, the computation time of potential flows around a circular cylinder decreases as the width of the Cartesian cell decreases (i.e., as the average number of target cells registered in the Cartesian cells decreases), resulting in a maximum speed up ratio of 200 [10]. The average number of fastest computations was 14.859, and if this number can be reduced to one, the speed up ratio may be a limiting factor. However, to reduce this average number, the computation time for registering the above algorithm increases immeasurably, as does the computer memory. Moreover, the number of elements in the dimensional array iSR(200,200,400) of the above program increases, even though much of the memory in the coarse cell regions of the target grid system is unused.

In the next section, a new algorithm is presented that overcomes these drawbacks.

3. New FORTRAN algorithm for maximum speed up

The most effective change to the FORTRAN algorithm is to change the order of the triple-do loops. In the previous algorithm, the do loop for the target cells (do k = 1, ic) was the innermost loop. In the modified algorithm, this loop is placed as the outermost loop, as shown in the program in Appendix III.

See how the above algorithm works. **Figure 3** shows an example of the target grid system of a circular cylinder (red) and a Cartesian grid system (black). The Cartesian cell (i = 9, j = 7) is shown enlarged on the right side of the figure. This cell covers or partially touches the target cell with ID numbers 13–16, 99–102, and 326–329.

Running the above algorithm gives the elements of the array iSR, as shown in **Table 1** (only the elements of Cartesian cell 9–7 are shown). The elements of parameters ir and k are in ascending order. However, the parameters i and j are not. To use this data more efficiently to search for the ID number of the target cell in which a target particle exists, the elements in the array iSR are rearranged using the algorithm in Appendix IV.

Table 2 lists the elements in the array iRec from i = 101 to i = 112, that were determined using the above algorithm. These elements are the ID numbers of the target cells (red cells in **Figure 3** on the right) in the Cartesian cell 9–7 (black square cell in **Figure 3** on the right), where the numbers i for the first and last cells are given as index (9,7,1) = 101 and index (9,7,2) = 112, respectively (**Table 3**).

Only the elements of the Cartesian cell 9–7 are shown.

Only, the elements of the Cartesian cell 9–7 are shown.

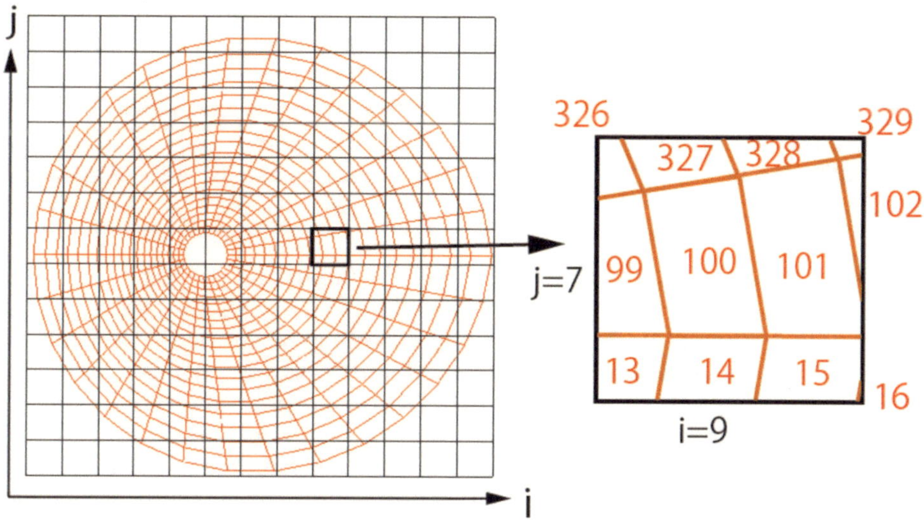

Figure 3.
Target grid cell system (red) and Cartesian grid cell system (black). The Cartesian cell 9–7 is magnified on the right side.

ir	**...**	**205**	**206**	**207**	**208**	**209**	**210**	**...**
iSR(ir,1) = k	...	13	14	15	16	16	17	...
iSR(ir,2) = i	...	9	9	9	9	10	10	...
iSR(ir,3) = j	...	7	7	7	7	7	7	...
400	**401**	**402**	**403**	**404**		**...**		
99	100	101	102	102		...		
9	9	9	9	10		...		
7	7	7	7	7		...		

Table 1.
Elements of array iSR.

i	**...**	**101**	**102**	**103**	**104**	**105**	
iRec(i)	...	13	14	15	16	99	
106	**107**	**108**	**109**	**110**	**111**	**112**	**...**
100	101	102	326	327	328	329	...

Table 2.
Elements of the array iRec (i) from i = 101 to 121.

index(9, 7, 1)	**101**
index(9, 7, 2)	112

Table 3.
Elements of the array index for the Cartesian cell 9–7.

If the x and y coordinates of the target particle are xp and yp, respectively, the ID number of the target cell in which this particle exists can be easily determined using the algorithm in Appendix V.

The subroutines "INorOUT" and "intersect" are shown in Appendices VI and VII. Whether a line AB and a line CD intersect and whether a point (x, y, x) exists in a square ABCD can be easily judged by comparing areas, as shown in the Appendices.

4. DSMC simulation with new FORTRAN algorithm

In a previous study [10], the CCR method was not used for the DSMC simulation, but for simple potential flows. In this section, dilute flows around a circular cylinder are simulated with DSMC using the new FORTRAN algorithm shown in Section 3. We consider two-dimensional flows of Argon around a circular cylinder with a radius 0.05 m and an inlet velocity of 2634.1 m/s. The parameters of the target and Cartesian grid cell systems are listed in **Tables 4** and **5**, respectively. The parameters for the dilute gas and DSMC simulations are listed in **Tables 6** and 7, respectively. To compare the results, the same parameters as in [11] were used for the gas and DSMC

Radius of the circular cylinder	**0.05 m**
Number of divisions in the radial direction	260
Number of divisions in the circumferential direction	462
Total grid cell count	120,120

Table 4.
Parameters of target grid cell system.

Maximum x-coordinate	**0.654 m**
Minimum x-coordinate	−0.396 m
Maximum y-coordinate	0.525 m
Minimum y-coordinate	−0.525 m
Number of cells in the x-direction	$1 \sim 2000$
Number of cells in the y-direction	$1 \sim 2000$

Table 5.
Parameters of Cartesian grid cell system.

Temperature of the cylinder surface	**500 K**
Stream temperature	200 K
Inlet velocity	2634.1 m/s
Number density	4.247×10^{20}
Molecular diameter	3.595×10^{-10} m
Molecular mass of Ar	6.63×10^{-26}kg

Table 6.
Parameters of dilute gas.

Time step	0.5×10^{-6}s
Total updates	40,000

Table 7.
Parameters for the DSMC simulation.

simulations. The full FORTRAN programs were written with double precision, compiled using the commercial compiler PGI 19.10–0 with an optimization level of four, and executed on a computer with an Intel Core i7-4770K processor with a clock speed of 3.50 GHz.

Figures 4 and **5** show the converged results for Mach number and temperature. Because the solutions were vertically symmetric, only the upper half is shown. These solutions agree well with the results in [11] (pp. 219 and 220).

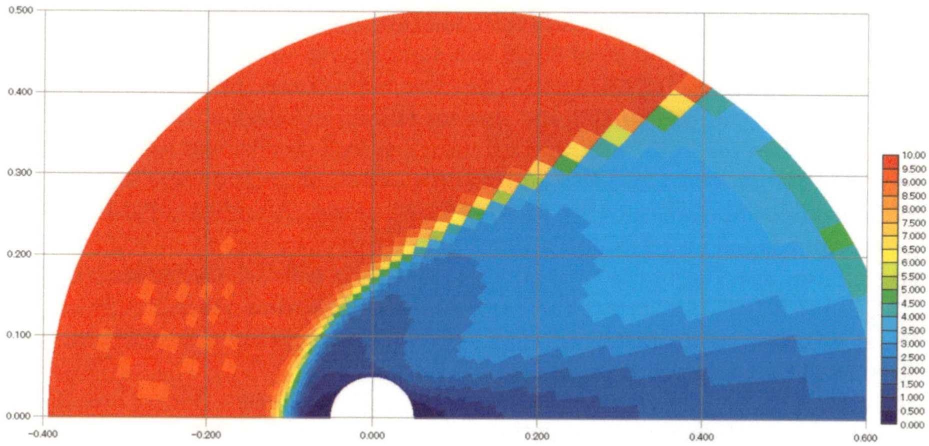

Figure 4.
Mach number. Only the upper half is displayed.

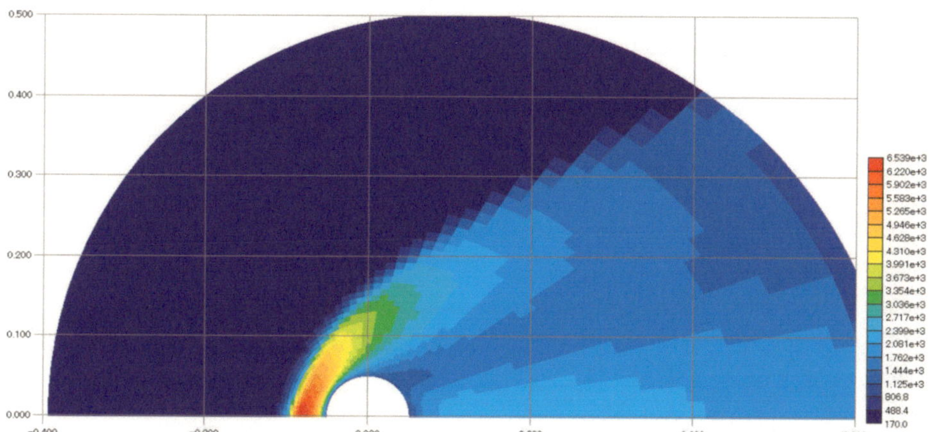

Figure 5.
Temperature. Only the upper half is displayed.

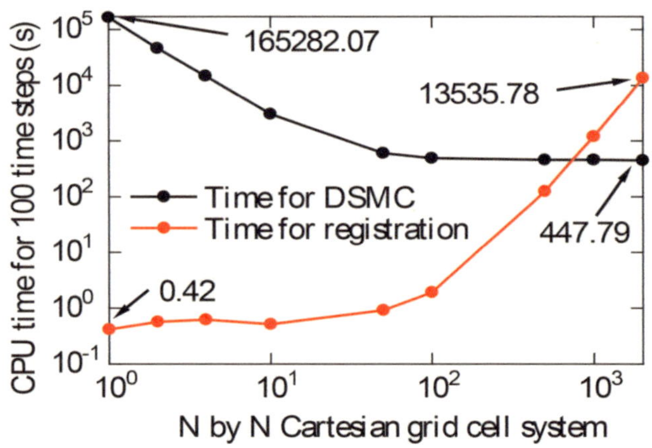

Figure 6.
Registration time and DSMC time for 100 time steps.

Figure 6 shows the CPU time for registration (red) and that for DSMC simulations (black) with different numbers of N (N × N Cartesian grid cell systems). For example, if a 1 × 1 Cartesian grid cell system, i.e., the brute-force approach, is used, the CPU time for registration is 0.42 s and that for DSMC simulation for 100 time steps is 165282.07 s. When the 2000 × 2000 Cartesian grid cell system was used, the CPU time for registration was 13535.78 s and that for DSMC simulation was 447.79 s. The DSMC speed up ratio was calculated to be 165282.07/447.79≈369.11. **Table 8** lists the CPU times and speed up ratios for N = 1–2000.

To study the speed up ratio for searching for target cells where the target particles exist, it is impractical to measure the CPU time for the search process because this process is mixed in the DSMC simulation. In addition, measuring the CPU time for all particles would affect total CPU time.

Instead, the average number of cells searched for one particle during one-time step was measured, and is summarized in **Table 9** and **Figures 7** and **8**. For example, if a 1 × 1 Cartesian grid cell system is used (the brute-force approach), this number is calculated to be 58208.83, which is nearly half of the total grid cell number of 120,120. Notably, this implies that the probability of any cell being searched is 50%.

N	Registration CPU time	DSMC CPU time	Speed up ratio
1	0.42	165282.07	1.00
2	0.57	46141.01	3.58
4	0.62	14822.28	11.15
10	0.51	3015.89	54.80
50	0.91	602.53	274.31
100	1.91	490.39	337.04
500	124.77	457.20	361.51
1000	1194.68	457.33	361.41
2000	13535.78	447.79	369.11

Table 8.
Registration time, DSMC time, and speed up ratio of DSMC calculation.

N	Averaged searched cell number for one particle during one-time step	Speed up ratio
1	58208.83	1.00
10	893.87	65.11
100	15.86	3670.44
300	3.95	14740.89
500	2.53	22989.85
800	1.87	31082.32
1000	1.64	35445.57
2000	1.32	44234.29

Table 9.
Averaged searched cell number for one particle during one-time step, and speed up ratio.

As the number of cells N of the Cartesian grid cell system increased, the number of searched cells decreases dramatically (**Table 9** and **Figure 7**), and the speed up ratio increased exponentially (**Table 9** and **Figure 8**). Most importantly, the CPU time for the DSMS reached a plateau at nearly N = 100 (**Figure 6**), indicating that N = 100 is sufficient for our simulation, although the speed up ratio for searching the target cells in **Figure 8** did not reach a plateau. This means that when N is more than 100, the CPU time for searching the target cells is negligible compared to the CPU time for the entire DSMC simulation. Consequently, the efficiency of our fast particle search algorithm reaches a maximum.

Moreover, as explained in Section 2, for the previous method, the average number of target cells registered in Cartesian cells for the fastest simulation was 14.859, but the average number of modified algorithms for N = 2000 was 1.59, excluding Cartesian cells with no registered target cells. This means that the new algorithm can reduce this average number (i.e., the size of the Cartesian cells can be reduced, and the Cartesian cell number can be increased) without sacrificing computational time or computer memory.

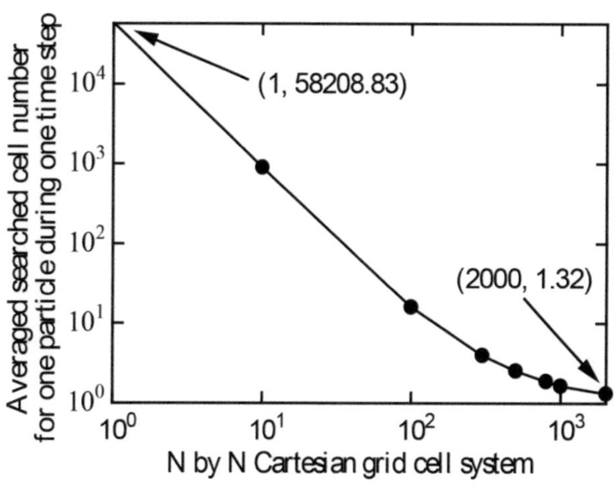

Figure 7.
Averaged searched cell number for one particle during one-time step.

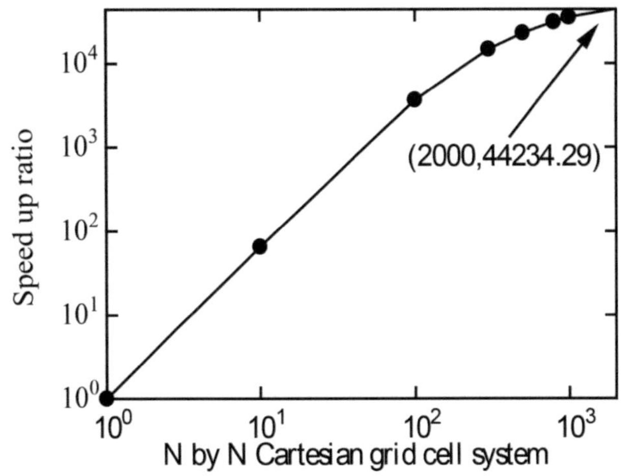

Figure 8.
Speed up ratio when searching cells.

5. Application for the measurement of multiple physical quantities

It was mentioned in the Introduction that particle searching and positioning are the most time-consuming processes in the direct simulation Monte Carlo (DSMC) method [1, 2] for dilute fluid flows, and in multiphase flow calculations [3–5]. As shown in Section 4, Cartesian cell registration (CCR) is an effective method for reducing the computational time required for DSMC calculations. In this section, we briefly explain how the CCR method can be used in fields of computational calculations other than DSMC.

We have worked on computational simulations for measuring multiple physical quantities using a single-motion sensor system [12]. **Figure 9a** shows a simple model

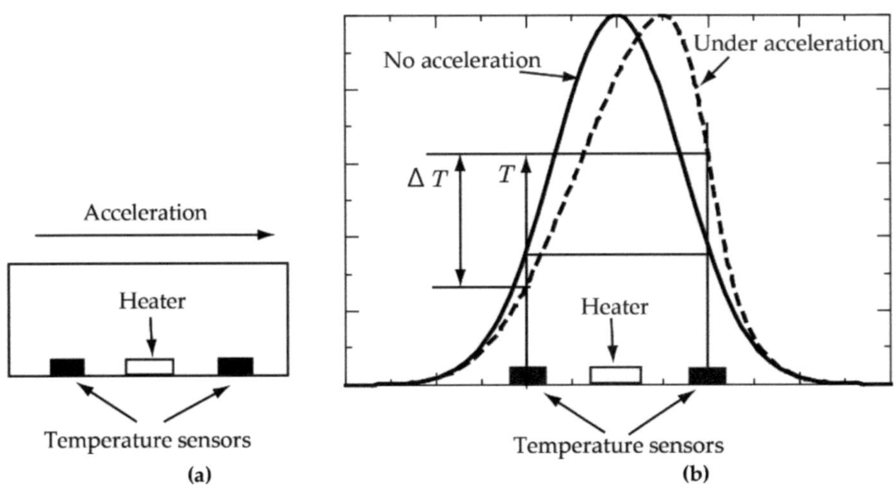

Figure 9.
Motion sensor system consisting of two temperature sensors and a heater placed in between (a) and temperature distributions with and without acceleration (b).

of a motion sensor system consisting of two temperature sensors and a heater placed in between. When this motion sensor was accelerated in the right direction, the top of the temperature distribution was inclined in the right direction owing to the buoyancy effect, as shown in **Figure 9b**. Analysis of a sufficient amount of data obtained by computer simulations showed that the magnitude of the acceleration is related to the temperature and temperature variations detected by the temperature sensors. Therefore, a motion sensor equipped with this data can measure acceleration due to temperature and the variation in temperature detected by the temperature sensors.

When this motion sensor is accelerated and rotated at the same time, the signal detected by the temperature sensors contains mixed information about both acceleration and rotation speed. The problem is to separate this signal into acceleration and rotation speeds. Using computer simulations, the two temperature outputs on the two temperature sensors can be calculated from the two inputs for acceleration and rotational speed. For example, when the acceleration is 10 m/s^2 and the rotational speed is 250°/s (black point A in **Figure 10a**), two temperature sensors in this motion sensor system detect 477.9 and 407.1 K (black point A' in (b)). Similarly, the rectangular region ABCD in (a) is converted to quadrilateral A'B'C'D' in (b). The remaining rectangular regions in (a) have been transformed into quadrilaterals, as shown in (b).

When a real motion sensor system installed in cars, drones, phones, game consoles, and virtual reality systems, for example, detects temperatures of 490 and 382 K with the simulation data in **Figure 9**, as indicated by the red circle in **Figure 10b**, acceleration and rotation speed can be estimated by inverting this red point in (b) to a red point in (a). The CCR method can be used to determine the quadrilateral A'B'C'D', in which the red point in (b) lies. That is, the red quadrilaterals in (b) are registered with black Cartesian cells also in (b). Then, the black Cartesian cell in which the red point exists can be calculated algebraically, and the quadrilateral A'B'C'D' is determined by the CCR method. Finally, even though the signal detected by the motion sensor contains mixed information about acceleration and rotation speed, as mentioned above, the corresponding rectangle ABCD in (a), where the inversely transformed red point is located, can be determined, and the acceleration and rotational speed can be easily calculated by our fast search algorithm.

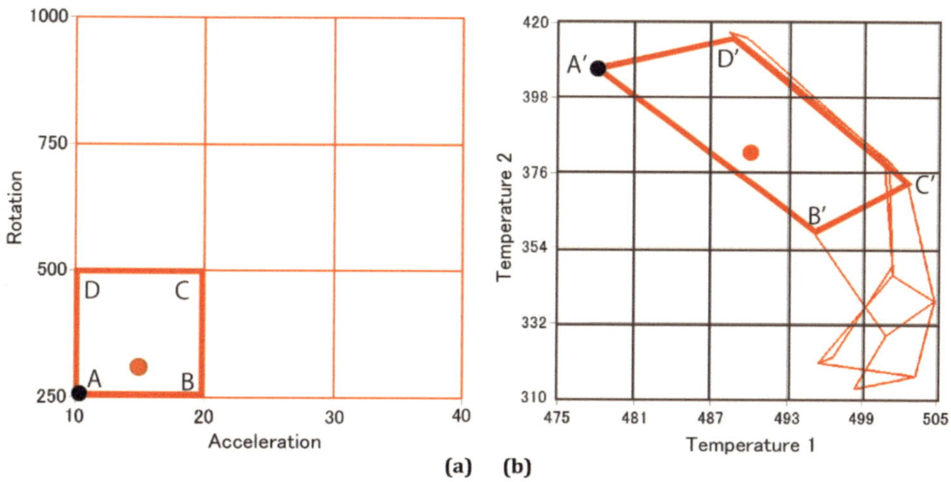

Figure 10.
Inputs of acceleration and rotation (a) and outputs of temperature (b). The region of input ABCD in (a) is transformed into the region of output A'B'C'D' in (b).

6. Conclusion

In this study, an efficient FORTRAN algorithm for the Cartesian cell registration (CCR) method is presented to further reduce both the computation time for registration and computer memory. Using this algorithm, the computational time for searching the target cell where the target particle is located was reduced by as much as 44,234 times. In addition, this algorithm was successfully applied to the DSMC simulation, resulting in a 369-fold reduction in computational time compared to the brute-force approach (searching all target cells until the target particle is present).

Future work will apply the new CCR method to multiphase flow calculations [3–5], the detection of static particle positions [13], and measurement of multiple physical quantities [12], as presented in Section 5. When measuring multiple physical quantities, the set of inputs and outputs must be a one-to-one correspondence. Removing this limitation will be our main task. We will also increase the number of simultaneously measured physical quantities from two to more and write our algorithm in Python, one of the most popular languages in engineering today, to achieve even faster processing.

Acknowledgements

The authors would like to thank Editage for English language editing.

Appendix I

```
Dimension cell(40,000,4,2), recent(200,200,2), iSR(200,200,400)

! The following arrays are x and y coordinates of four corners of target cell with ID number k.
! cell(k,1,1), cell(k,1,2)
! cell(k,2,1), cell(k,2,2)
! cell(k,3,1), cell(k,3,2)
! cell(k,4,1), cell(k,4,2)
! The maximum cell number 40000 is tentative.
! The order of the four corners is clockwise or counterclockwise.

! The following arrays are x and y coordinates of the centers of Cartesian cell with ID number
! i and j.
! recent(i,j,1), recent(i,j,2)
! The maximum ID number 200 is tentative.
! i is the ID number of the Cartesian cell in x direction
! j is the ID number of the Cartesian cell in y direction

! iSR(i,j,1) is the number of registered target cells in the Cartesian cell ij.
! iSR(i,j,2 ~ 400) are ID numbers of the target cells which are registered in the Cartesian cell ij.
! 400 (tentative) is the maximum number of the target cells minus 1 which Cartesian cell ij
! can memorize.
! Initialize iSR(i,j,1); the number of registered target cells in the Cartesian cell ij.
      do i = 1,nRecx! nRecx is the maximum number of i
      do j = 1,nRecy! nRecy is the maximum number of j
iSR(i,j,1) = 0
      enddo
      enddo
```

```
! Start registration
   do i = 1,nRecx
   do j = 1,nRecy
   do k = 1,ic       ! do loop for the target cells
                     ! ic is the total number of the target cells
! The center of the target cell
xjc = (cell(k,1,1) + cell(k,2,1) + cell(k,3,1) + cell(k,4,1))/4
yjc = (cell(k,1,2) + cell(k,2,2) + cell(k,3,2) + cell(k,4,2))/4

! The center of the rectangle cell
      xc = recent (i,j,1)
yc = recent (i,j,2)
! The distance between the centers
      al = dsqrt((xc-xjc)**2 + (yc-yjc)**2)
          if(al.le.ak*dx) then! ak is arbitrary number. dx is width of Cartesian cell
          iii = iSR(i,j,1)
          iii = iii + 1
          iSR(i,j,1) = iii! iii is the total number of registered target cells in the Cartesian cell ij
          iSR(i,j,iii + 1) = k! k is the ID number of the target cell
          endif
enddo
enddo
enddo

! End of registration
```

Appendix II

```
! ID numbers of the Cartesian cell i and j
i = (xp-xRecmin)/dx + 1! xRecmin is the minimum x coordinate of the Cartesian grid system
   j = (yp-yRecmin)/dx + 1! yRecmin is the minimum y coordinate of the Cartesian grid system

! The total number of target cells registered in Cartesian cell ij
iscn = iSR(i,j,1)

! Find out the target cell where the target particle exists
   do k = 1,iscn
   ii = iSR(i,j,k + 1)! candidate ID number of the target cell
   x1 = cell(ii,1,1)! x coordinate of the first corner on the target cell
   y1 = cell(ii,1,2)! y coordinate of the first corner on the target cell
   x2 = cell(ii,2,1)! x coordinate of the second corner on the target cell
y2 = cell(ii,2,2)! y coordinate of the second corner on the target cell
   x3 = cell(ii,3,1)! x coordinate of the third corner on the target cell
   y3 = cell(ii,3,2)! y coordinate of the third corner on the target cell
   x4 = cell(ii,4,1)! x coordinate of the fourth corner on the target cell
   y4 = cell(ii,4,2)! y coordinate of the fourth corner on the target cell

call INorOUT(x1,y1,x2,y2,x3,y3,x4,y4,xp,yp,iflag)! subroutine to investigate
                                  ! whether the target particle exists in this target cell.
                                  ! This is shown in Appendix VI.
    if(iflag.eq.0) then! the target particle exists in this target cell
    goto 999
    endif
enddo

999 continue! the ID number (the element of parameter k) has been found.
```

Appendix III

```
Dimension iSR(100,000,3), rec(100,100,4,2), cell(100,000,4,2)
Dimension index(100,100,2), iRec(100000)
! The element numbers 100,000 and 100 are tentative. These depend on the total cell number
! of the target grid system and the Cartesian grid system

! Initialize iSR

      do i = 1,100,000
      iSR(i,1) = 0
iSR(i,2) = 0
iSR(i,3) = 0
enddo

ir = 0! ir is the total number of registered set of k, i and j

! The following arrays are x and y coordinates of four corners of target cell with ID number k.
!   cell(k,1,1), cell(k,1,2)! the first corner
!   cell(k,2,1), cell(k,2,2)! the second corner
!   cell(k,3,1), cell(k,3,2)! the third corner
!   cell(k,4,1), cell(k,4,2)! the forth corner

! do loop for the target cells is the outermost loop
   do k = 1,isc0    ! do loop for the target cells
                    ! isc0 is the total number of the target cells

! The maximum and minimum values of the four corners of the target cells
   xmax2 = −1.0d10
if(xmax2.lt.cell(k,1,1)) xmax2 = cell(k,1,1)
if(xmax2.lt.cell(k,2,1)) xmax2 = cell(k,2,1)
if(xmax2.lt.cell(k,3,1)) xmax2 = cell(k,3,1)
if(xmax2.lt.cell(k,4,1)) xmax2 = cell(k,4,1)

      xmin2 = 1.0d10
if(xmin2.gt.cell(k,1,1)) xmin2 = cell(k,1,1)
if(xmin2.gt.cell(k,2,1)) xmin2 = cell(k,2,1)
if(xmin2.gt.cell(k,3,1)) xmin2 = cell(k,3,1)
if(xmin2.gt.cell(k,4,1)) xmin2 = cell(k,4,1)

      ymax2 = −1.0d10
if(ymax2.lt.cell(k,1,2)) ymax2 = cell(k,1,2)
if(ymax2.lt.cell(k,2,2)) ymax2 = cell(k,2,2)
if(ymax2.lt.cell(k,3,2)) ymax2 = cell(k,3,2)
if(ymax2.lt.cell(k,4,2)) ymax2 = cell(k,4,2)

      ymin2 = 1.0d10
if(ymin2.gt.cell(k,1,2)) ymin2 = cell(k,1,2)
if(ymin2.gt.cell(k,2,2)) ymin2 = cell(k,2,2)
if(ymin2.gt.cell(k,3,2)) ymin2 = cell(k,3,2)
if(ymin2.gt.cell(k,4,2)) ymin2 = cell(k,4,2)

! Find out in which Cartesian cell these corners exist

   is = (xmin2-xRecmin)/dx + 1! xRecmin is the minimum x coordinate of the Cartesian grid
ie = (xmax2-xRecmin)/dx + 1

js = (ymin2-yRecmin)/dx + 1! yRecmin is the minimum y coordinate of the Cartesian grid
   je = (ymax2-yRecmin)/dx + 1
```

```
! Start registration
! do loops for the Cartesian cells are inner loops
do i = is,ie
  do j = js,je

do ii = 1,4
  Ax = cell(k,ii,1)
  Ay = cell(k,ii,2)
  iii = ii + 1
if(ii.eq.4) iii = 1
  Bx = cell(k,iii,1)
  By = cell(k,iii,2)
! The following arrays are x and y coordinates of the four corners of Cartesian cell ij.
!   rec(i,j,1,1), rec(i,j,1,2)! the first corner
!   rec(i,j,2,1), rec(i,j,2,2)! the second corner
!   rec(i,j,3,1), rec(i,j,3,2)! the third corner
!   rec(i,j,4,1), rec(i,j,4,2)! the fourth corner

do jj = 1,4
  Cx = rec(i,j,jj,1)
Cy = rec(i,j,jj,2)
jjj = jj + 1
if(jj.eq.4) jjj = 1
  Ex = rec(i,j,jjj,1)
Ey = rec(i,j,jjj,2)

! See if the side of the target cell (AB) and the side of the Cartesian cell (CE) intersect

call intersect(Ax,Ay,Bx,By,Cx,Cy,Ex,Ey,iflag,err)! this subroutine is shown in Appendix VII.

if(iflag.eq.0) goto 555! if they intersect, go to 555
                          ! and register the target cell k with the Cartesian cell ij
enddo
enddo

! if no sides of the target cell (AB) and of the Cartesian cell (CD) intersect,
! then see if the whole region of target cell k is inside the Cartesian cell ij.

xA = rec(i,j,1,1)
  yA = rec(i,j,1,2)
  xB = rec(i,j,2,1)
  yB = rec(i,j,2,2)
xC = rec(i,j,3,1)
yC = rec(i,j,3,2)
xD = rec(i,j,4,1)
yD = rec(i,j,4,2)

  ifs = 0
do lll = 1,4
  x = Scell(k,lll,1)
  y = Scell(k,lll,2)
  call INorOUT(xA,yA,xB,yB,xC,yC,xD,yD,x,y,iflag)
  ifs = ifs + iflag
  enddo
if(ifs.eq.0) goto 555! if the whole region of target cell k is inside the Cartesian cell ij, go to 555
                          ! and register the target cell k with the Cartesian cell ij
! Or inversely see if the whole region of the Cartesian cell ij is inside the target cell k

    Ax = cell(k,1,1)
```

```
      Ay = cell(k,1,2)
      Bx = cell(k,2,1)
      By = cell(k,2,2)
    Cx = cell(k,3,1)
      Cy = cell(k,3,2)
      Ex = cell(k,4,1)
    Ey = cell(k,4,2)

      ifs = 0
    do lll = 1,4
      x = rec(i,j,lll,1)
      y = rec(i,j,lll,2)
      call INorOUT(Ax,Ay,Bx,By,Cx,Cy,Ex,Ey,x,y,iflag)

      ifs = ifs + iflag

  enddo

  if(ifs.eq.0) goto 555! if the whole region of the Cartesian cell ij is inside the target cell k,
                        ! go to 555 and register the target cell k with the Cartesian cell ij

  goto 666! if there is nothing to register, go to 666.

  555 continue

  ir = ir + 1

  iSR(ir,1) = k! id number of the target cell
  iSR(ir,2) = i! id number of the Cartesian cell in x direction
  iSR(ir,3) = j! id number of the Cartesian cell in y direction

  666 continue

  enddo
  enddo
  enddo

  ! end of registration
```

Appendix IV

```
  ! Rearrange the ID number of target cells in the order of Cartesian cell i and j
        in = 0

        do i = 1,nRecx! nRecx is the maximum number of i
        do j = 1,nRecy! nRecy is the maximum number of j
        is = in
        do k = 1,ir! ir is the total number of registered sets of k, i and j

        if(iSR(k,2).eq.i.and.iSR(k,3).eq.j) then
        in = in+1
      iRec(in) = iSR(k,1)! the ID number of the target cell k registered in Cartesian cell ij is
                         ! put into array iRec of ID number in
        endif
      enddo
```

```
! ID numbers of target cells registered in Cartesian cell ij starts from index(i,j,1)-th element
! to index(i,j,2)-th element in arrary iRec(in). Namely, in = index(i,j,1) ~ index(i,j,2)

                index(i,j,1) = is+1
                index(i,j,2) = in

            enddo
            enddo
! End of rearrangement
```

Appendix V

```
! ID numbers of the Cartesian cell i and j
i = (xp-xRecmin)/dx + 1! xRecmin is the minimum x coordinate of the Cartesian grid
     j = (yp-yRecmin)/dx + 1! yRecmin is the minimum y coordinate of the Cartesian grid

is = index(i,j,1)
ie = index(i,j,2)

! Find out the target cell where the target particle exists
   do k = is, ie
   ii = iRec(k)! ID number of target cell registered in Cartesian cell ij.
   x1 = cell(ii,1,1)! x coordinate of the first corner on the target cell
   y1 = cell(ii,1,2)! y coordinate of the first corner on the target cell
   x2 = cell(ii,2,1)! x coordinate of the second corner on the target cell
   y2 = cell(ii,2,2)! y coordinate of the second corner on the target cell
   x3 = cell(ii,3,1)! x coordinate of the third corner on the target cell
   y3 = cell(ii,3,2)! y coordinate of the third corner on the target cell
   x4 = cell(ii,4,1)! x coordinate of the fourth corner on the target cell
   y4 = cell(ii,4,2)! y coordinate of the fourth corner on the target cell

call INorOUT(x1,y1,x2,y2,x3,y3,x4,y4,xp,yp,iflag)! subroutine to investigate
                                     ! whether the target particle exists in this target cell
    if(iflag.eq.0) then! the target particle exists in this target cell
    goto 999
    endif
enddo

999 continue! the ID number is the element of parameter k
```

Appendix VI

```
*************************************************************
      subroutine INorOUT(xA,yA,xB,yB,xC,yC,xD,yD,x,y,iflag)
*************************************************************
! Check if a point (x,y,z) exists in a square ABCD
! The order of the four corners ABCD is clockwise or counterclockwise.
! Just comparing areas shows if a point (x,y,z) exists in a square ABCD or not.

implicit real*8 (a-h,o-z)

      call rectangle(xA,yA,xB,yB,xC,yC,xD,yD,area)
```

```
      call triangle(xA,yA,xB,yB,x,y,area1)
      call triangle(xB,yB,xC,yC,x,y,area2)
      call triangle(xC,yC,xD,yD,x,y,area3)
      call triangle(xD,yD,xA,yA,x,y,area4)

      er = area-area1-area2-area3-area4

iflag = 1! false
   e1 = dabs(er/area)
   if(e1.lt.2.5d-9) iflag = 0! The threshold value should be adjusted depending on the cell size
if(area.eq.0.0d0) iflag = 1

   return
   end
*************************************************************
      subroutine rectangle(xA,yA,xB,yB,xC,yC,xD,yD,area)
*************************************************************
      implicit real*8 (a-h,o-z)

area1 = dabs((xA-xB)*(yC-yA)-(xC-xA)*(yA-yB))/2
      area2 = dabs((xA-xD)*(yC-yA)-(xC-xA)*(yA-yD))/2

      area = area1 + area2
      return
   end
```

Appendix VII

```
*******************************************************************
subroutine intersect(Ax,Ay,Bx,By,Cx,Cy,Dx,Dy,iflag,err)
*******************************************************************
*   Check if line AB and line CD intersect.
*   Just comparing areas shows if line AB and line CD intersect or not.
*
      implicit real*8 (a-h,o-z)
iflag = 1
      call triangle(Ax,Ay,Cx,Cy,Dx,Dy,ACD)
      call triangle(Bx,By,Cx,Cy,Dx,Dy,BCD)
      call triangle(Cx,Cy,Ax,Ay,Bx,By,CAB)
      call triangle(Dx,Dy,Ax,Ay,Bx,By,DAB)

      err = dabs(ACD + BCD-CAB-DAB)/(ACD + BCD + CAB+DAB)

   if(err.lt.1.0d-10) iflag = 0! The threshold value should be adjusted depending on the cell size
   return
   end
*******************************************************************
subroutine triangle(xA,yA,xB,yB,xC,yC,area)
*******************************************************************
      implicit real*8 (a-h,o-z)

! Area of triangle ABC
      area = dabs((xA-xB)*(yC-yA)-(xC-xA)*(yA-yB))/2

      return
      end
```

Author details

Yoshifumi Ogami* and Siddique Kamran
Ritsumeikan University, Kusatsu-City, Japan

*Address all correspondence to: ogami@se.ritsumei.ac.jp

References

[1] Bird GA. Molecular Gas Dynamics and the Direct Simulation of Gas Flows. Oxford: Oxford University Press; 1994

[2] Bird GA. Direct simulation and the Boltzmann equation. Physics of Fluids. 1970;**13**:2676-2681

[3] Shojaee S, Hosseini SH, Razavi BS. Computational fluid dynamics simulation of multiphase flow in structured packings. Journal of Applied Mathematics. 2012;**2012**:1-17

[4] Florice NM, Andrei K. Modelling and simulation of multiphase flow applicable to processes in oil and gas industry. Chemical Product and Process Modeling. 2019;**20170066**:1-16. DOI: 10.1515/cppm-2017-0066

[5] Parsi M, Kara M, Agrawal M, Kesana N, Jatale A, et al. CFD simulation of sand particle erosion under multiphase flow conditions. Wear. 2017; **376-377**(B):1176-1184

[6] Blazek J. Principles of grid generation. In: Computational Fluid Dynamics: Principles and Applications. 3rd ed. Amsterdam: Elsevier; 2015

[7] Liang J, Yan C, Du BQ. An algorithm study of three dimensional DSMC simulation based on two-level Cartesian coordinates grid structure. Acta Aerodynamica Sinica. 2010;**28**:466-471

[8] Wang Z, Li L, Zhang B, Liu HBCP. Particle positioning techniques for DSMC method. Journal of Aeronautics, Astronautics and Aviation. 2019;**51**: 225-236

[9] Wang C, Cheng J, Ji L, Lu Y, Sun Y, et al. 2-D DSMC algorithm based on Delaunay triangles. Journal of Tsinghua University (Science and Technology). 2015;**10**:1079-1086. DOI: 10.16511/j.cnki.qhdxxb.2015.22.010

[10] Ogami Y. Fast algorithms for particle searching and positioning by cell registration and area comparison. Trends in Computer Science and Information Technology. 2021;**6**:7-16

[11] Bird GA. The DSMC Method. Version 1.2. 2013. Available from: http://www.gab.com.au/index.html.

[12] Siddique K, Ogami Y. Computational study on thermal motion sensors that can measure acceleration and rotation simultaneously. Sensors. 2022;**22**(18): 6744. DOI: 10.3390/s22186744 [Accessed: April 17, 2023]

[13] Thapa S, Lukut N, Selhuber-Unkel C, Cherstvy AG, Metzler R. Transient superdiffusion of polydisperse vacuoles in highly motile amoeboid cells. The Journal of Chemical Physics. 2019;**150**: 144901-1-144901-18

Chapter 8

Reduction of Kernel Mechanism Processing Time in Multimedia Authoring Tool

Marvin Chandra Wijaya

Abstract

Multimedia authoring tools is a tool for designing multimedia documents based on multimedia presentation scenarios designed by the user. Visual media (images, video, animation, text, etc.) and non-visual media (sound, music, etc.) are combined into a multimedia document. A multimedia programming language represents multimedia documents. The core process in multimedia authoring tools is called the kernel mechanism, which consists of two mandatory processes: modeling and verification. Verification carried out consists of time-computation and spatiotemporal verification. The time needed for the verification process on a multimedia authoring tool is highly dependent on the object or media. The relationship between processing time and the amount of media is exponential, so if a lot of media is processed, it will take a very long time. This is the main obstacle of a multimedia authoring tool at this time. The incrementality algorithm can be applied to the verification process in order to reduce processing time to verify multimedia documents. Based on the experimental results, the processing time can be reduced by up to 75%.

Keywords: multimedia authoring tools, multimedia authoring attributes, multimedia authoring modeling, incrementality, verification

1. Introduction

Multimedia authoring tools are tools that function to combine various kinds of media into a multimedia document. The multimedia document is formed from a combination of several media according to the temporal and spatial layout [1–3]. Multimedia authoring tools must have good performance to help users produce proper multimedia documents [4]. The resulting document is represented using a multimedia language, which currently uses several multimedia languages. A multimedia authoring tool performs a kernel function in merging various independent media into a multimedia document. In general, there are three main parts in a multimedia authoring tool: the process of entering temporal data and spatial layouts, kernel mechanisms, and multimedia documents. The three parts are executed sequentially, as shown in **Figure 1**.

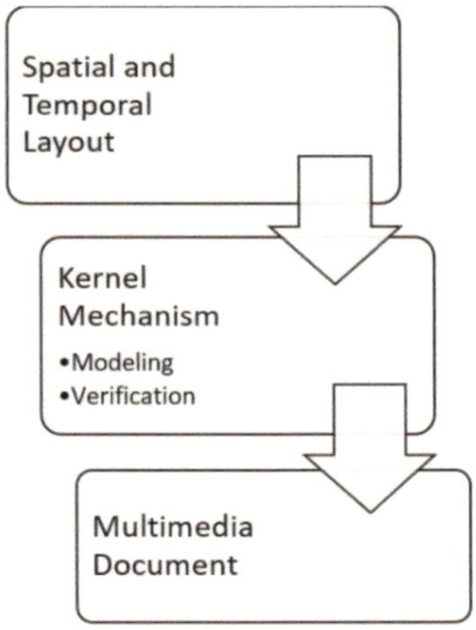

Figure 1.
Three steps in multimedia authoring tools [5].

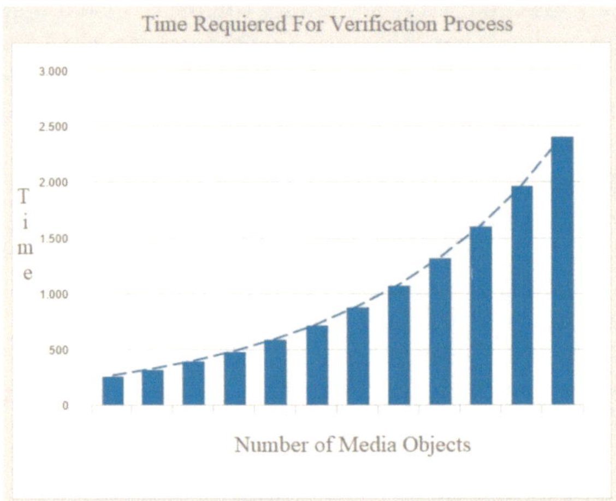

Figure 2.
Relation between kernel processing time with the number of media objects [6].

One of the problems in the process of producing multimedia documents using multimedia authoring tools is the time required for the kernel process. The kernel processing time depends on the number of media inside the multimedia document, as shown in **Figure 2** [6].

The processing time occurs because many processes occur in the kernel mechanism. Therefore, making the kernel mechanism process more efficient in terms of the runtime process is challenging.

2. Multimedia authoring attributes

A multimedia authoring tool has several attributes that inform that the multimedia authoring tool is functioning properly [7]. The attributes are service attribute, edit attribute, performance attribute, and formal model, as shown in **Figure 3**.

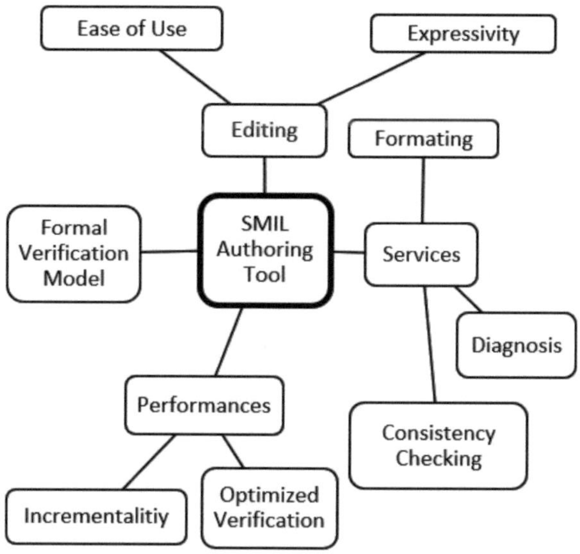

Figure 3.
Multimedia authoring attributes [8].

Attribute editing is a feature provided for the user in designing multimedia documents. There are two factors that become parameters in editing attributes: ease of use and expression. The editing process is the first part of the kernel mechanism in a multimedia authoring tool. Multimedia authoring tools have editing attributes tools including the editing process, user interface, syntax errors, error messages box, information box, and others. The more facilities in the multimedia authoring tool, the better the editing attribute value will be. Ease of use is a factor that underlies the author's ease in designing multimedia documents [9]. One of the ease-of-use factors in the multimedia authoring tool is the user interface for the authoring proses, which makes it easy to use [10]. **Figure 4** is an example of a multimedia authoring tool that uses WYSIWYG as a user interface.

The services provided by a multimedia authoring tool can be measured indirectly by investigating the following:

- Formulate and find questions or goals.
- Search for data
- Conduct a preliminary study
- Data processing
- Data output.

Figure 5 shows the hierarchy of document multimedia.

Performance attribute is a direct or indirect assessment of the performance results of the multimedia authoring tool. There are two parameters in the performance attribute: incrementality and optimized verification. Incrementality is the ability to process data by considering changes in input data that occur [11]. Software with the

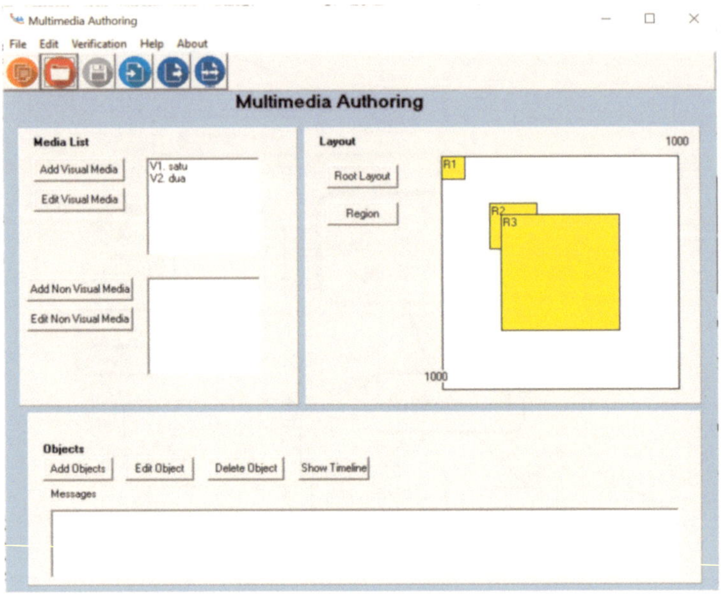

Figure 4.
Example of WYSIWYG interface on a multimedia authoring tool [5].

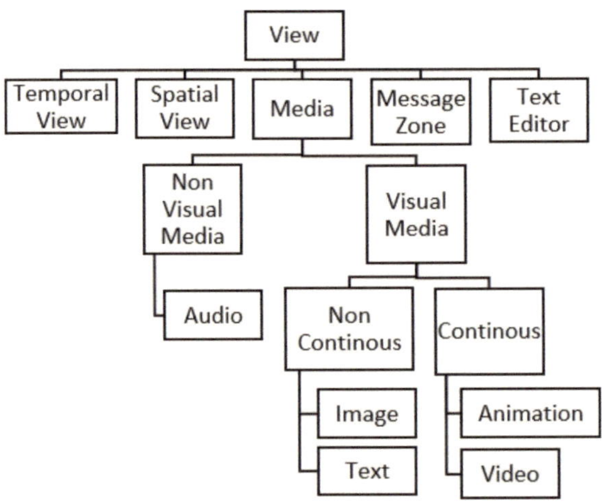

Figure 5.
Hierarchy of document multimedia [8].

incrementality concept is considering the process of recalculating based on the changed input data.

Verification is important in the multimedia authoring tool to maintain the integrity of multimedia documents. Several strategies to verify multimedia documents include analysis of editing process, time verification, spatial verification, and other validation.

Formal verification is a process to check a system that has the correct functioning of the system. Formal verification is useful to check a system in software in producing multimedia documents [12, 13].

3. Multimedia programming languages

There are several multimedia programming languages, such as synchronized multimedia integration language (SMIL), nested context language (NCL), hypertext markup language 5 (HTML 5), and others. Two programming languages that are quite widely used are SMIL and NCL.

Multimedia documents can be represented using a markup language, called SMIL (Synchronized Multimedia Integration Language) [14]. SMIL is a recommendation from the World Wide Web Consortium (www.w3c.org) for multimedia programming languages and part of XML [15, 16]. SMIL is a markup language similar to HTML and a XML variant. SMIL can control multimedia presentations, for example regarding on-screen layout and timing. The SMIL structure is defined by the module structure, consisting of <smil>, <head>, and < body> elements. The SMIL structure is as follows:

```
<smil>
    <head>
    ...meta information, layout, title, transistion effect.
    </head>
    <body>
    ... media, group, linking, animation.
    </body>
</smil>
```

Nested context language (also called as NCL) is an XML-based declarative language that represents multimedia documents. NCL has a modular structure that combines modules into a language profile [17, 18]. NCL has the ability to create multimedia presentations by creating application content on various existing devices. NCL as well as SMIL can also determine the temporal and spatial space for the media to be played. NCL has very strict separation rules between content and structure. This causes NCL not to define the media content itself, so NCL must define the glue that holds the media together and arranges them in a spatial and temporal form to become a complete multimedia document [19]. Structure of nested context language (NCL) is as follows:

```
<?xml?>
<ncl>
<head>
      <region Base>
....
</region Base>
<descriptor Base>
....
</descriptor Base>
<connector Base>
....
</connector Base>
</head>
<body>
    ..............
</body>
</ncl>
```

4. Kernel mechanism

There are two essential processes in the kernel mechanism: the verification and modeling processes [20]. The verification process is time-consuming in a multimedia authoring tool process. Therefore, modifying the verification process and using an incrementality algorithm is necessary to improve its performance [5]. Modifications in the kernel mechanism are shown in **Figure 6**.

The verification process in the kernel mechanism is added with an incrementality algorithm to streamline its performance. The incrementality process will reduce processing time in the kernel mechanism of a multimedia authoring tool [18].

Incrementality of time calculation verification is combined with spatiotemporal verification process. This can make the verification process more efficient. The time needed in the verification process will be better. The verification process is running as usual in its original state, and there has been no change whatsoever. The incremental algorithm will be applied when there is a change or modification process. The verification process occurs for each object grouped by region in a multimedia presentation. If an error is found, the system will inform the user. Eq. (1) is used to find conflicts that occur in a region.

$$\forall Media_a, \forall Media_b, a < m, b < m, \exists End(media_a) < Begin(media_b) \tag{1}$$

Where:
$Media_a$ = Media Object a

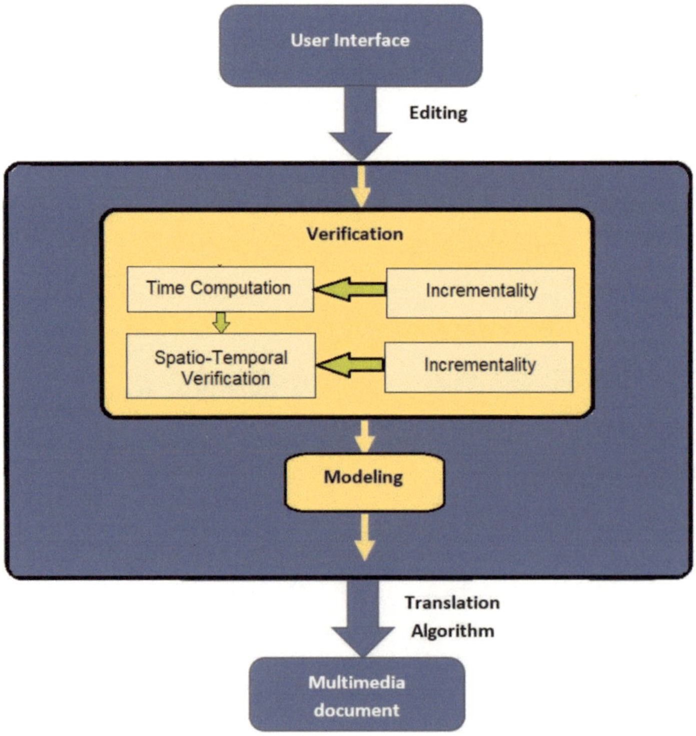

Figure 6.
Modified kernel mechanism.

$Media_b$ = Media Object b
m = Number of media objects
Begin = Begin time
End = End time.

The incrementality algorithm of the time computation verification process is as follows:

---

1. Prepare each object for each area
2. Convert the timeline form from two dimensions to three dimensions.
3. The process of setting the media based on time
4. For each area {
5. If there is a change in a media {
6. For each medium {
7. IF Media_end(i) < Media_begin (i) then give message
8. End for
9. end if

Spatialtemporal verification is used to verify objects in different overlapping regions at the same time interval. Multimedia documents will be checked first to determine whether each region is overlapping: spatial and whole overlapping. Any possible overlap in each pair of regions will be included in a list. When the list is complete, all media on the list are checked for broadcast times. The verification process is applied to two overlapping media in different regions. In the incrementality of the spatiotemporal verification algorithm, verification will be applied to objects that the user has changed.

Boolean Eq. (2) is used to find two overlapping regions.

$$((Lb > La)\&\&(Lb < La + Wa)||(Lb + Wb > La)\&\&(Lb + Wb < La + Wa)\&\&$$
$$((Tb > Ta)\&\&(Tb < Ta + Ha))||(Ta + Ha > Ta)\&\&(Tb + Hb < Ta + Ha)) \quad (2)$$

Where:
La = left position of media a.
Lb = left position of media b.
Ta = top position of media a.
Tb = top position of media b.
Wa = width of object a.
Wb = width of object b.
Ha = height of object a.
Hb = height of object b.

The incrementality algorithm of the spatiotemporal verification process is as follows:

---

1. Prepare the substrate
2. Convert the timeline from two dimensions to three dimensions
3. Organize media by time
4. For regions
5. Process the overlapping areas
6. Ends For
7. For overlapping areas
8. If there is a change in the media
9. For main region{
10. For the next region{
11. If the start of the main region < the start of the second region then submit a message
12. End For
13. End For
14. End If
15. End For

The algorithm was tested using the SMIL language to determine its success. A multimedia presentation scenario consisted of four regional layouts and eight video media. The eight video media are displayed in a predetermined region, because the number of media is more than the number of regions, and a combined parallel and sequential scenario is designed. The scenario is designed so that all media can be displayed correctly and can be tested using the designed incrementality algorithm. The following is a multimedia document in the form of the SMIL language which is used for testing the incrementality algorithm [5].

```
<smil>.
<head>.
    <layout>.
    <root-layout  height="110"  width="110" />.
    <region  width="480"  height="480"  top="10"  left="10"
    id="area1" />.
```

```
<region width="480" height="480" top="10" left="510"
id="area2" />.
<region width="480" height="480" top="510" left="10"
id="area3" />.
<region width="480" height="480" top="510" left="510"
id="area4" />.
</layout>.
</head>.
<body>.
<seq>.
<video region="areal" src="movie5.vid" id="id1" dur="160"/>.
<par dur="160 s">.
<video region="area2" src="movie6.vid" id="id2" dur="160"/>.
<video region="area3" src="movie7.vid" id="id3" dur="160"/>.
</par>.
<par dur="160 s">.
<video region="area2" src="movie8.vid" id="id4" dur="160"/>.
<seq dur="160 s">.
<video region="area3" src="movid9.vid" id="id5" dur="80"/>.
<video region="area4" src="movie10.vid" id="id6" dur="80"/>.
</seq>.
</par>.
<par dur="160 s">.
<video region="area2" src="movie11.vid" id="id7" dur="160"/>.
<video region="area3" src="movie12.vid" id="id8" dur="160"/>.
<video region="area3" src="movie13.vid" id="id9" dur="160"/>.
 </par>.
 </seq>.
</body>.
</smil>.
```

When the duration of the "movie5.vid" is changed from 160 to 180, the following process is

- Process: Dur1(par) = 160

  Dur2(par) = 180

- Process 2: D1 (par) = D2 (par)

- Process 3:"par" is invariant

Where:
D1 = initial duration
D2 = modification duration

The verification process will get a value of D = 160 which is regulated in the coding program, but a different value is obtained after the verification process. This causes the algorithm to declare an error. The author will also receive information on the error message in the message box on the user interface.

Number of Media	Number of Region	Number media per region	Time Required (with Incrementality)	Time Required (without Incrementality)
36	12	3	6 ms	31 ms
36	3	12	9 ms	69 ms
72	12	6	20 ms	129 ms
72	6	12	19 ms	137 ms
96	12	8	33 ms	180 ms
96	8	12	30 ms	174 ms
120	12	10	47 ms	238 ms
120	10	12	45 ms	229 ms
144	12	12	55 ms	297 ms

Table 1.
Process time comparison.

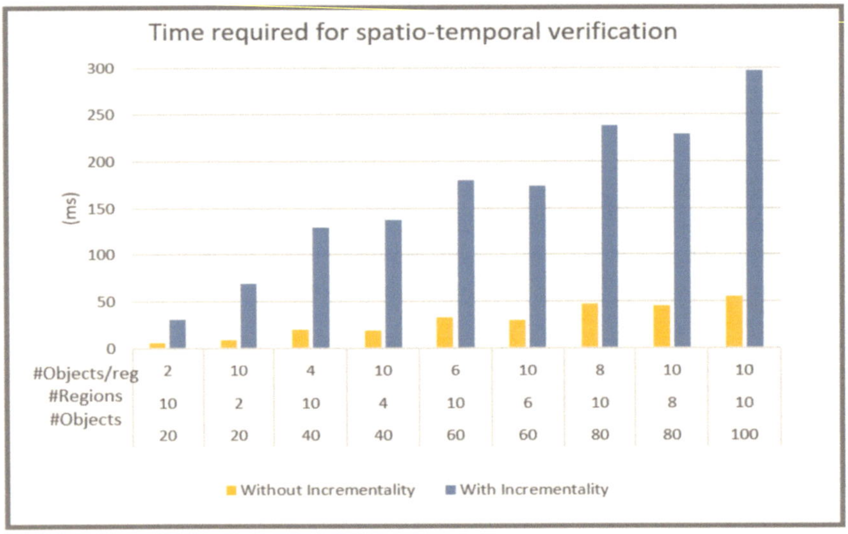

Figure 7.
Process time comparison graph.

Testing the incrementality algorithm in the verification process is done by comparing the time of the verification process. The comparison being tested is an algorithm that uses incrementality and without incrementality. **Table 1** and **Figure** 7 show a comparison of the time needed to calculate the time using and without the incrementality algorithm in the verification process. Based on the experimental results, it can be seen that the incrementality algorithm has succeeded in reducing the processing time by almost 75%.

5. Conclusions

Multimedia authoring tools combine and process visual media (images, videos, animations, text, and others) and non-visual media (sound, music, and others) into a multimedia document. Currently, there are several languages for multimedia programming, such as SMIL, NCL, HTML 5, and others. The core process in the multimedia authoring tool is called the kernel mechanism, which consists of two mandatory processes, namely modeling and verification. There have been many verifications and modeling developments in multimedia authoring tools. Multimedia authoring tool has several attributes such as editing, formatting, services, formal verification model, and performance. All attributes must be met by a multimedia authoring tool to produce efficient multimedia documents, errors, and others.

Based on the experimental results, it was found that adding media to a document would result in a longer verification process, both time computation and spatiotemporal verification. The incrementality algorithm can reduce processing time to verify multimedia documents. Based on the experimental results, the processing time can be reduced by up to 75%.

Acknowledgements

I also thank Departement of Computer Engineering, Faculty of Engineering, Maranatha Christian University, Indonesia, for the support in writing this book chapter.

Conflict of interest

The authors declare no conflict of interest.

Author details

Marvin Chandra Wijaya
Maranatha Christian University, Bandung, Indonesia

*Address all correspondence to: marvin.cw@eng.maranatha.edu

References

[1] De Mattos DP, Muchaluat-Saade DC. Steve: A hypermedia authoring tool based on the simple interactive multimedia model. Proceedings of the ACM Symposium on Document Engineering. 2018;**14**(4):1-24. DOI: 10.1145/3209280.3209521

[2] Belhadad Y, Refoufi A, Roose P. Spatial reasoning about multimedia document for a profile based adaptation: Combining distances, directions and topologies. Multimedia Tools and Applications. 2018;**77**(23):30437-30474. DOI: 10.1007/s11042-018-6080-8

[3] Huang B, Yi S, Chan WT. Spatio-temporal information integration in XML. Future Generation Computer Systems. 2004;**20**(7):1157-1170. DOI: 10.1016/j.future.2003.11.005

[4] Picinin D, Farines JM, Santos CAS, Koliver C. A design-oriented method to build correct hypermedia documents. Multimedia Tools and Applications. 2018;**77**(16):21003-21032. DOI: 10.1007/s11042-017-5325-2

[5] Wijaya MC, Maksom Z, Abdullah MHL. Speedup multimedia authoring process: Incrementality of multimedia authoring verification. Journal of Engineering Science and Technology Review. 2022;**15**(3):170-176. DOI: 10.25103/jestr.153.19

[6] Wijaya MC, Maksom Z, Abdullah MHL. Novel framework for translation algorithms in multimedia authoring tools. IAENG International Journal of Computer Science. 2022;**49**(3):628-636 [Online]. Available: http://www.iaeng.org/IJCS/issues_v49/issue_3/IJCS_49_3_02.pdf

[7] Bouyakoub S, Belkhir A. SMIL builder: An incremental authoring tool for SMIL documents. ACM Transactions on Multimedia Computing, Communications, and Applications. 2011;**7**(1):1-30. DOI: 10.1145/1870121.1870123

[8] Wijaya MC, Maksom Z, Abdullah MHL. A brief of review: Multimedia authoring tool attributes. Ingénierie des Systèmes d'Information. 2021;**26**(1):1-11. DOI: 10.18280/isi.260101

[9] Pulungan FF, Sudiharto DW, Brotoharsono T. Pattern lock and the app based on context, ease of use aspect in comparison. Indones. Journal of Electrical and Engineering Informatics. 2019;**7**(1):58-66. DOI: 10.11591/ijeei.v7i1.936

[10] Yang CC, Yang YZ. SMILAuthor: An authoring system for SMIL-based multimedia presentations. Multimedia Tools and Applications. 2003;**21**(3):243-260. DOI: 10.1023/A:1025770817293

[11] Carlsson M. Monads for incremental computing. ACM SIGPLAN Notices. 2002;**37**(9):26-35. DOI: 10.1145/583852.581482

[12] Belkhir A, Bouyakoub-Smail S. Formal design of smil documents. In: Proceeding of Third International Conference on Web Information Systems and Technologies (Webist 2007); 3-6 March 2007; Spain. 2007. pp. 396-399. DOI: 10.5220/0001271403960399

[13] Mekahlia FZ, Ghomari A, Yazid S, Djenouri D. Temporal and spatial coherence verification in SMIL documents with Hoare logic and disjunctive constraints: A hybrid formal method. Journal of Integrated Design

and Process Science. 2017;**20**(3):39-70. DOI: 10.3233/jid-2016-0020

[14] Meixner B. Hypervideos and interactive multimedia presentations. ACM Computing Surveys. 2017;**50**(1): 1-34. DOI: 10.1145/3038925

[15] Antonacci MJ, Muchaluat-Saade DC, Rodrigues RF, Soares LFG. Improving the expressiveness of XML-based hypermedia authoring languages. Multimedia Modeling: Modeling Multimedia Information and Systems, MMM. 2000;**2020**:71-88

[16] Hoschka P. The application/smil and application/smil+xml media types. RFC. 2006;**4536**:1-8. DOI: 10.17487/RFC4536

[17] Yovine S, Olivero A, Córdoba L. An approach for the verification of the temporal consistency of NCL applications. In: Proceeding of Simposio Brasileiro de Sistemas Multimídia e Web (WebMedia'10); 5-8 October 2010; Brazil. 2010. pp. 179-184

[18] Júnior DP, Farines JM, Koliver C. An approach to verify live NCL applications. In: Proceedings of the 2012 Brazilian Symposium on Multimedia and the Web (WebMedia'12); 15-18 October 2012; Brazil. 2012. pp. 223-231. DOI: 10.1145/2382636.2382685

[19] Ratamero EM. Tutorial sobre a linguagem de programação NCL (Nested Context Language). Universidade Federal Fluminense; 2007. p. 11

[20] Bouyakoub S, Belkhir A. A Spatio-temporal authoring tool for multimedia SMIL documents. International Journal of Multimedia and Technology. 2012; **2**(3):83-88

Chapter 9

Improving the Relevance, Speed, and Computational Efficiency of Semantic Search through Database Indexing: A Review

Xiaofeng Yang

Abstract

In a large dataset, finding top most relevant matches to fulfill search queries and similar application becomes an increasing challenge. Facing this challenge, researchers and engineers has done extensive work in both regime of algorithm design and system implementation to create practical solutions that fuel powerful tools. Techniques like caching, search index, fuzzy indexing has been widely adopted in both SQL and XML data models, then provided to the database user, the application programmer, as a product. In this chapter, we review data indexing algorithms that efficiently compute semantic search query results, and discuss systems that implement them. We compare the pros and cons of each algorithm, in terms of time and space complexity, as well as implementation difficulty. We also analyze trade-offs in real-world system design examples, in hope to provide the readers more insights when they want to efficiently find an answer through semantic search query, or even build their own application using these databases.

Keywords: algorithms, application of algorithms, database systems, web development, relevance improvement

1. Introduction

Interacting with the web is becoming an essential part of our life, all of us at some point, are web application users: we search for spelling of a word, purchase groceries and clothes from online catalog, play video contents over the phone or on TV, and check our bank statements through our online account. Behind all these applications, there are data storage and retrieval tasks performed many times to power our query, and present accurate results in a timely manner. In this ever increasing scale of web data, finding top most relevant matches becomes increasingly challenging, both to the application provider such as Wayfair and Nordstorm, and to the infrastructure provider that sells computational resource and platforms, such as Amazon AWS. Facing this challenge, researchers and engineers has done extensive work to create practical solutions. Products like SQL databases, NoSQL databases, techniques like caching, search index, have been widely adopted and provided to the application programmer as a product. Web application developers often want to know, out of all these technologies, which tech stack shall we pick to accomplish a task? How do we evaluate the strength and weakness of each option, and pick the solution that best fit our situation?

This is a hard question and often requires some trial and error, along with some collaboration work that involves expertise from multiple areas. However, there are some perspectives that many developers and designers in the industry have came up with, and proven effective in engineering and business challenges. We try to provide some tool and guidelines to our reader, to help build a framework to reason about the choices of these technologies.

We will first discuss data models behind popular web applications, introduce how data is being stored and used, and go through example products. We then discuss a few poplar techniques used in each product to optimize system performance, their concept and history, and how to evaluate the effectiveness of them. We then walk through a special optimization use case in web semantic search, and show how application engineer solves a real problem starting from analyzing the use case.

2. Data models

Data models defines how application programmers reason about the problem, and is usually the first step in designing a web application using databases.

To group databases into two high level categories by their data model and design principals, SQL (Structured Query Language) [1] and NoSQL (Not Only SQL) [2] are two very high-level types of database management systems.

2.1 SQL databases

The family of SQL databases has been around since the early days of web applications, and is usually associated with the query language SQL. A typical SQL database is based on a relational data model, where data is organized into tables with predefined columns and rows. SQL languages are used to query these table. Here is a simple SQL language example query:

```
SELECT column1, column2, ... FROM example_table;
```

What this query does is to retrieve data stored under the name of column1 and column2, in a table called example table. On top of that, we can specify filtering condition of the data, order by which the data is sorted, or even recursively specify query in the result of another query.

SQL databases have a fixed schema, that dictates the structure of the data, as well as the relationships between tables. Any changes to the schema require modifications to the database schema. This sometimes means additional migration work to an application programmer using a SQL database, they will have to back-fill the new database instance with new schema, which can be time-consuming and error prone [3].

The reason of having a fixed schema is behind the common use cases when SQL databases are first designed: we want them to be ACID-compliant, which means that they guarantee the properties of atomicity, consistency, isolation, and durability for all transactions [4]. These properties help build foundational assumptions the application programmer will rely on: they knows if part of their write to the database succeeded, the full body of this write must have succeeded (atomicity); The database will not give clients contradictory query answers (consistency); The behavior one client interacting with the database will not be affected by other clients the database serves (isolation), and once data is persisted into the database, it will loyally stay there for later usage (durability). These properties have helped application developers to build software for bank transaction without worrying about trapping into a partial result state, like client A seeing the money transferred over while client B not seeing it. However, these properties are usually built based on carefully designed mechanisms such as locks, and can be show to response, and expensive in terms of computational resources required.

2.2 NoSQL databases

The word NoSQL was originally created to describe not-only-SQL representable data models [5]. NoSQL databases therefore cover a large variety of different data models, such as key-value store [6], document-based data model [7], column-family database [8], or graph database [9].

Many NoSQL databases have a flexible schema comparing to SQL databases, allowing for changes to be made to the data model without requiring modifications to the database schema. Adding an additional level of key-value recursively or adding a paragraph in the document can be performed online while the database is still serving incoming traffic. This property has positioned NoSQL databases in great advantage for fast-changing, always-online web applications: in the scenarios when we anticipate frequent data schema change, the adaptation of NoSQL data model takes a significant maintenance burden away from application engineers.

As oppose to SQL databases, which use SQL, a standardized language to manage/manipulate relational data as their query language, NoSQL databases use a variety of query languages that are specific to their data model. These languages may require different syntax and logic than SQL [10].

Why did people propose NoSQL databases? In what scenarios do they show advantage over SQL databases?

One answer comes from their scalability: SQL databases are typically vertically scalable, meaning that they can only handle larger amounts of data by adding more resources to a single server. This is limiting the scale that a SQL database handles, since the complexity and cost of building one single powerful computer grows fast, and is limited by modern computer architecture. On the contrary, NoSQL databases are usually horizontally scalable, meaning that they can handle larger amounts of data by adding more servers to a distributed system. This means as the number of users and workload grows, business owners using a NoSQL database can easily grow their capability by buying more commodity hardware, such as regular computers you can obtain from Amazon, and add them to their database deployment.

As we previously discussed, SQL databases are designed to be ACID-compliant. NoSQL databases make case-by-case sacrifice on some of these properties, for improved scalability or performance.

As a result of all these different design considerations, SQL databases are usually the first choice for applications that perform sophisticated transactions on structured data, while NoSQL databases shine in the domain of remaining applications that require scalability, flexibility, and handling of unstructured or semi-structured data.

3. Case study

In this section we look at real commercial database systems and their usage, to better understand the trade-offs made by engineers.

3.1 SQL database

One well-known SQL DB is MySQL, a popular open-source relational database management system (RDBMS) that is owned by Oracle Corporation. Over the years, it has been widely used by application programmers as a backend database for both web and enterprise applications. MySQL is a popular choice because of its reliable performance, scalability, and ease of use, and can be found under the hood of web-based applications such as online photo editing services, content management systems such as library software, and many small to medium sized e-commerce websites.

Let us consider a simple web application that allows users to register, log in, and post comments on a website. MySQL can be used as the database management system to store and retrieve user information and comments.

3.1.1 User registration

- When a user registers on the website, their registration information (such as username, email, and password) is collected through an HTML form through client side UI (user interface).

- Upon receiving the form, the web application backend takes all these information and sends a SQL query to the database to insert a new row into the "users" table. The query might look like this:

```
INSERT INTO users (username, email, password) VALUES
    ('ExampleUser', 'user@example.com', 'hashed_password');
```

where the password is encrypted and not stored in plain text. Each user will only have one password, so each username will only have one record in the table. This means, we can use the column username to identify a unique row in the table. We call this username column a primary key of the table.

- After executing the above write query, the database stores this user's registration information securely as a row in the users table.

3.1.2 User login

- When a user tries to log in, our web application asks for their login credentials (username and password) and compares with our own record, to verify that this user is the same person who owns the account.

- Upon receiving the HTML form, the application backend sends a SQL query to the database to check if the provided username and password match the row previously stored in the database. The query might look like this:

```
SELECT * FROM users WHERE username='ExampleUser' AND password='hashed_password';
```

- The database executes the above read query, and returns the result to the web application backend for comparison.

- Based on the comparison result, the web application grants or denies access to the user.

3.1.3 Posting comments

- After logging in, now the user can post a comment on our page.
- Again through HTML form in the UI, the web application collects the comment content along with the logged in user's ID.
- It sends an SQL query to the database to insert a new row into the "comments" table. The query might look like this:

```
INSERT INTO comments (user_name, content) VALUES
    ('ExampleUser', 'This is a great article!');
```

- The MySQL database stores the comment along with the user's username, linking the comment to the specific user in the users table. This usage of username across two tables is called a foreign key.

3.1.4 Retrieving comments

At the time when the web application displays comments on a page, it sends an SQL query to the database to fetch the comments for that page. The query might look like this:

```
SELECT * FROM comments WHERE page_id=1,000,001;
```

- The database executes the read query and returns the comments to the web application backend.
- The web application backend then sends the retrieved data to its front-end, and then display the comments on the page.

In the above example, read and write operations are performed on the database. MySQL, as a common SQL database can provide these capabilities, to store and manage user information and comments. The web application interacts with the database by sending SQL queries and retrieving the results, allowing users to register, log in, post comments, and view existing comments on the website.

3.2 NoSQL database: Dynamo DB

We study Amazon Dynamo DB as an example NoSQL database. Amazon DynamoDB is a fully managed, proprietary NoSQL database service offered by Amazon Web Services (AWS). It was first introduced by Amazon in 2012 and is designed to provide low-latency, scalable and highly available database solutions for modern applications.

DynamoDB is a document-oriented, key-value store database, which means that data is stored as key-value pairs or as structured JSON documents.

3.2.1 Performance

As a highly scalable database, DynamoDB can handle petabyte-scale workloads with automatic scaling of throughput capacity, and it provides a rich set of data consistency levels, durability and security features. However, it does not guarantee full ACID (Atomicity, Consistency, Isolation, Durability) properties due to their simplified data model and focus on high-performance and scalability.

In a key-value store, each key-value pair is typically treated as an independent entity, and transactions spanning multiple key-value pairs are not natively supported. Updates to individual key-value pairs are usually atomic, but there is no built-in mechanism to ensure atomicity across multiple key-value pairs.

DynamoDB also prioritize high-performance and scalability by allowing concurrent access to data without extensive locking mechanisms. This can lead to potential data conflicts or race conditions when multiple transactions access and modify the same keys simultaneously.

3.2.2 Application

DynamoDB is optimized for low-latency and high-throughput access patterns, making it suitable for applications that require fast and predictable performance, such as gaming, e-commerce, social media, and IoT applications.

DynamoDB is offered as a fully managed service, meaning that AWS takes care of the infrastructure, scaling, backups, and maintenance of the database, allowing developers to focus on building their applications.

A serverless application is an architecture and computing model where the developer focuses on writing code for specific functions or services, without the need to manage the underlying infrastructure or servers. In a serverless model, the cloud provider takes care of provisioning, scaling, and managing the servers, allowing developers to focus solely on writing and deploying code [11]. DynamoDB also integrates with other AWS services such as AWS Lambda, Amazon Kinesis, and Amazon S3 [12, 13], making it easy to build serverless applications with a variety of use cases.

DynamoDB also offers a global table feature, which allows for automatic replication of tables across multiple AWS regions. This enables low-latency access to data for users distributed worldwide, as well as fault tolerance when machines in some region suffers from unavoidable disasters.

Integrating with AWS using Dynamo DB offers benefits such as reduced infrastructure management, automatic scalability, cost optimization, and rapid development cycles. DynamoDB becomes a popular database service among AWS customers due to its scalability, performance, and ease of use.

4. General optimization techniques

The readers may start to wonder, under the hood of these commercial databases, what algorithms are they using to provide these reduced cost, improved speed and reliability? Two most popular optimization techniques are indexing and caching, and they are widely adopted in traditional RDBMS (relational database management systems), in many variations of forms. However, the simple and powerful main idea is not changing: instead of computing the results every time a query is executed, pre-computed partial or full results are stored and served.

However, these approaches differ on what to store, where to store the partial results, and how to make sure that the stored (cached) data is up-to-date. Different algorithms are proposed to find the best index to store at a certain use case, so that the following goals are achieved: (1) the result is not stale according to the applications' requirement, and can be directly used by the web application without affecting its correctness; (2) The cost of storing such results is acceptable; (3) For a typical workload in the use case, storing these additional results is effective.

In this section, we will go over details of both indexing and caching and understand how they provide optimization.

4.1 Indexing

4.1.1 Concept and history

An index is a data structure that allows for faster data retrieval by enabling the database to locate and access specific rows more quickly. The concept of database indexing was first introduced in the early days of computer databases, as a means to improve data retrieval efficiency. The initial development of indexing dates back to the 1960s when the first database management systems (DBMS) were being developed.

One of the earliest known instances of indexing in databases can be traced back to IBM's System R project in the 1970s [14]. System R, a research project at IBM's San Jose Research Laboratory, focused on the development of a relational database system. System R introduced the concept of indexing to improve the performance of relational database queries.

In 1970, Edgar F. Codd published his groundbreaking paper titled "A Relational Model of Data for Large Shared Data Banks" [15] which laid the foundation for the relational database model. Although the paper did not explicitly mention indexing, it proposed the idea of a primary key as a means to uniquely identify records in a table, which essentially served as an implicit form of indexing.

Over the years, indexing techniques and algorithms have evolved, and different types of indexes, such as B-trees, hash indexes, and bitmap indexes, have been developed to suit various data access patterns and requirements.

It's important to note that while the concept of indexing emerged in the 1960s and 1970s, the specific techniques and algorithms for indexing have continued to be refined and optimized, driven by advancements in computer hardware, storage systems, and database technologies.

4.1.2 Indexing in database

Here are some key reasons why indexes are created in a modern database:

- Improved Query Performance. Indexes provide a way to locate data in a database quickly. By creating an index on one or more columns, the RDBMS can use the index to narrow down the search space and retrieve relevant data more efficiently. This results in faster query execution times, especially when working with large tables or when executing complex queries involving joins or filtering conditions. Instead of scanning the entire table, the database engine can use the index to locate specific rows or a range of rows, reducing disk I/O and improving data retrieval speed. Well-designed indexing can reduce the latency of such queries by order of magnitudes.

- Efficient Data Sorting. Indexes can also be utilized to speed up sorting operations. When a query requires sorting the result set based on a specific column, an index on that column allows the RDBMS to retrieve the data in the desired order more efficiently, eliminating the need for a separate sorting step.

- Constraint Enforcement. Indexes can enforce the uniqueness of values in a column or set of columns by creating unique indexes. Unique indexes prevent duplicate values from being inserted into the indexed columns, maintaining data integrity and supporting the enforcement of primary key or unique constraints.

It's worth noting that while indexes improve query performance, they also have some trade-offs. Indexes require additional disk space to store the index data structure, and they can introduce overhead during data modification operations (such as inserts, updates, and deletes) as the indexes need to be updated along with the actual data. In a SQL database where atomicity is required, this usually means locking all the data to modify, resulting into an increased latency in the write operation. It's essential to carefully design and maintain indexes based on the specific usage patterns, query workload, and data characteristics of the database to achieve optimal performance improvements while considering the associated costs and trade-offs.

4.1.3 Design of Indexing

Here are some of the most popular form of indexing storing different data:

- B-tree Indexing. B-trees are commonly used in databases to implement indexing. B-trees are balanced tree structures that allow for fast searches, insertions, and deletions of data. B-trees are optimized for disk-based storage and can be used to index large amounts of data efficiently.

- Hash Indexing. Hash indexing uses a hash function to map the values of a column to a specific location in the index. This allows for fast lookups and is commonly used in in-memory databases.

- Bitmap Indexing. Bitmap indexing is used to index columns that have a low cardinality (i.e., a small number of distinct values), we call these data *sparse*. In these cases, storing the data in a bitmap of special values is a more compact representation than storing data in its original format. Bitmap index uses a bitmap to represent each unique value in the column, and each bit in the bitmap represents a row in the table. This allows for fast querying of data, especially for columns that have a small number of distinct values.

- Full-Text Indexing. Full-text indexing is used to index text-based data, such as documents or articles. Full-text indexing uses techniques such as tokenization and stemming to create an index of words or phrases that appear in the text. This allows for efficient searching of text-based data.

4.2 Caching

4.2.1 Concept and history

The concept of caching is not only limited in database context: Caching refers to the process of storing frequently accessed or computed data in a temporary storage location that allows for faster retrieval. In a database, this usually means storing the full or intermediate query result in the database system's memory (or fast access layer), allowing direct presentation of these results for the same or similar queries in the future. Caching aims to improve system performance, reduce latency, and optimize resource utilization.

The cached data is typically kept closer to the requesting entity (such as an application or a user) to minimize the latency associated with accessing the original source or performing expensive computations repeatedly. We can apply caching on the client side of the application, which is closer than the application backend; We can also apply caching on the first server that serves client traffic, which is closer than the database server.

We can cache various types of data, including database query results, web pages, images, API responses, computed values, file contents. The choice of what to cache depends on the specific use case and the data that is accessed frequently or requires significant processing time. If the application requires frequent access of the same file, then we should consider caching that file content. If multiple operations in the application rely on the same computed values, we should consider caching that computed value for repeated usage.

4.2.2 Caching in databases

When talking about caching in databases, we usually mean to cache query results or partial results, for repeated usage in a typical workload where the application performs query on the database. Caching in the database can be categorized into Buffer Caching and Query Result Caching:

- Buffer Caching. Buffer caching, also known as disk block caching or page caching, was introduced to speed up disk I/O operations in database systems. It involved caching disk blocks or database pages in memory to reduce the need for physical disk reads and writes. This technique was widely adopted in database management systems (DBMS) in the 1970s and 1980s.

- Query Result Caching. The concept of query result caching, where the results of frequently executed queries are cached for faster retrieval, gained prominence in the early 2000s with the development of advanced database systems. It became more prevalent as databases evolved to handle larger data volumes and more complex queries.

Modern database management systems, both relational and non-relational, have incorporated various forms of caching to optimize performance. Caching techniques include result set caching, query plan caching, and page buffer caching. These caching mechanisms leverage the memory resources of the database server or distributed cache clusters to store frequently accessed data and query results, minimizing the need for costly disk I/O operations and reducing query execution times.

4.2.3 Design of caching

Caching effectiveness depends on factors such as cache size, cache eviction policies, and the access patterns of the data. An optimized caching strategy tailored to the specific use case and data access patterns is crucial to maximizing the benefits of caching and achieving optimal performance improvements.

Most of the caching effectiveness depends on the spatial and temporal locality of the data access patterns. For example:

- LRU (Least Recently Use) Eviction Policy exploits the temporal locality of the access pattern by removing the least recently used item to make room for new entries. For improving the data structure efficiency, usually a Clock algorithm is used to simulate LRU with less memory overhead.

- Low Inter-reference Recency Set (LIRS) policy improves LRU by better handling the scan workload such that when access pattern starts scanning the data set, the access does not kick the "real" hot elements from the cache (scan-resistence).

- Pre-fetching is another caching strategy that proactively loads the data near the accessed data set. It exploits the fact that data close to each other are more likely to be accessed together, hence after pre-fetching, the data to be accessed next is more likely already been paged into the cache.

4.3 Metrics

Metrics provide useful tools for engineers to understand the current performance of a system, and allow reasoning about its strength and weakness, then find a path towards more optimal designs. Latency, throughput and availability are three important metrics when we evaluate a database system.

4.3.1 Latency

In a database system, latency refers to the time delay or the amount of time it takes for a query or operation to be executed and for the result to be returned. It is a measure of the response time or the delay experienced by the user or application when interacting with the database.

Latency in a database system can be caused by various factors, including:

- Network. The time taken for data to travel between the client and the database server over the network can contribute to latency. Network latency depends on factors such as the distance between the client and server, network congestion, and the quality of the network infrastructure.

- Disk I/O. When accessing data stored on disk, the time required to read or write data from or to the storage medium can impact latency. Disk I/O latency depends on factors such as disk speed, seek times, and the number of concurrent I/O operations.

- Query Processing and Execution. The time taken by the database server to process and execute a query can contribute to latency. This includes parsing the query, optimizing the query execution plan, accessing the necessary data from storage, and performing any required computations or joins.

To measure latency in a database system, many database systems offer monitoring and performance analysis tools that provide insights into query execution times, wait times, I/O statistics, and other performance-related metrics. These tools can help identify bottlenecks and measure latency at various stages of query execution.

Caching reduces latency by storing frequently accessed data closer to the requesting entity. When data is already cached, it can be quickly retrieved from the cache, eliminating the need to access the original data source, which may involve disk I/O, network latency, or complex computations. By minimizing the time required to retrieve data, caching reduces the overall response time and improves latency.

4.3.2 Throughput

Throughput refers to the amount of work or the number of operations that can be processed within a given time period. It represents the system's capacity to handle a certain volume of requests or transactions per unit of time.

Throughput is often measured in terms of transactions per second (TPS), queries per second (QPS), or operations per second (OPS). It reflects the system's ability to efficiently process and execute database operations, such as reads, writes, updates, and queries.

To measure throughput in a database system, we can consider the following approaches:

- Workload Testing: Perform workload testing by simulating a realistic or expected workload on the database system. This involves executing a set of representative queries or operations, with a specified concurrency level, and measuring the number of successful transactions or operations completed within a specific time frame. Workload testing tools or frameworks can help automate and streamline this process.

- Performance Monitoring and Alerting: Use performance monitoring tools or features provided by the database system to collect metrics related to throughput. These tools often provide information about the number of queries executed, transactions processed, or operations completed per unit of time. By monitoring and analyzing these metrics, you can gain insights into the system's throughput capacity and performance.

- Load Testing: Conduct load testing by subjecting the database system to a higher-than-usual workload, gradually increasing the load to observe how the system handles it. Load testing tools can generate a high volume of concurrent requests or transactions to stress-test the system and measure its throughput under varying load conditions.

Caching enhances throughput by reducing the load on the underlying data storage or computation resources. When data is cached, subsequent requests for the same data can be served directly from the cache, without putting additional strain on the database system. By avoiding redundant and expensive operations, such as disk I/O or complex computations, caching frees up resources to handle other requests and increases the overall throughput capacity of the system.

Another way to look at this is, caching offloads the database system by reducing the number of queries or requests that need to be processed. As frequently accessed data is stored in the cache, the number of requests hitting the database decreases. This reduction in the database load allows the system to handle a higher volume of requests and improves the overall scalability of the database.

4.3.3 Availability

Database availability refers to the ability of a database system to remain accessible and operational, providing uninterrupted access to data and services. It ensures that users or applications can reliably access and interact with the database system whenever they need to, without experiencing significant downtime or disruptions.

High availability is a critical aspect of database systems, particularly in scenarios where downtime can lead to business disruptions, financial losses, or degraded user experiences. Database availability is often expressed as a percentage of uptime, such as "five nines availability" (99.999%), indicating a highly available system with very minimal downtime. Achieving high availability requires careful planning, robust infrastructure, redundancy, fault-tolerant design, and continuous monitoring and maintenance practices.

Key factors that contribute to database availability include:

- System Redundancy: Implementing redundant components, such as multiple database servers, network connections, and storage systems, can help mitigate the impact of hardware failures or other infrastructure issues. Redundancy ensures that if one component fails, the system can seamlessly switch to a backup or standby component to maintain availability.

- Fault Tolerance: Building fault-tolerant mechanisms into the database system helps it withstand and recover from failures gracefully. This includes features like automatic failover, replication, and clustering, which enable the system to continue operating even if individual components or nodes encounter issues.

- Disaster Recovery: Having robust disaster recovery plans and mechanisms in place is essential for maintaining database availability during catastrophic events or major disruptions. This involves implementing off-site backups, replication to remote locations, and procedures for recovering the system and data in case of a disaster.

- Monitoring and Alerting: Proactive monitoring and alerting systems help identify potential issues or performance degradation in real-time. Monitoring tools can track various metrics related to system health, performance, and availability, enabling administrators to detect and address issues promptly to minimize downtime.

- Maintenance and Upgrades: Database systems require regular maintenance, patching, and upgrades. Ensuring availability during these activities is crucial. Techniques such as rolling upgrades, where components are upgraded one at a time without interrupting overall system operation, can help maintain availability during maintenance.

Caching may also contribute to availability by providing a fallback option when the original data source is temporarily unavailable or experiencing performance issues. If the cache holds a copy of the requested data, it can continue serving that data even if the underlying database is unavailable. This helps ensure uninterrupted access to critical data and improves the overall availability of the system.

5. Optimization in special applications - most relevant search

5.1 Turning use case to requirements

In the sections above, we have built our basic tools to understand database performance. However, in the ever growing problem space of web application, new use cases derive from complex, ever-changing ways human interact with a computer in modern applications. Imagine the scenario of a web user performing a semantic search on the internet. Instead of asking performance questions using the system language that we just acquired, they may inspect their own user experience and then ask the following questions:

- Can I get the results faster if I do not mind out-of-date results as long as they are accurate?
- What if I do not mind if a small percentage of results are not accurate?
- Can I review the results before committing to running a full query?
- Can I get the most relevant results and then decide if I want to continue the query?

Unlike well-defined metrics, these are case-specific performance requirements raised by a user. Application developers have to consider these requirements before they proceed to picking the technology to build a system. We call this step a use case analysis, which usually results into a product requirement.

A use case is a description or representation of a specific interaction or scenario that demonstrates how a system or software is used to accomplish a particular goal or provide value to its users. It outlines the steps, actions, and interactions between actors (users or external systems) and the system being analyzed or designed.

Use cases analysis is commonly required in software development, requirements engineering, and system analysis to capture and communicate functional requirements and system behavior. They help define and understand the various ways in which users or external entities interact with the system to achieve specific objectives.

To better meet the requirements for this use case, we need to first translate these questions the user raised into languages of system performance, build an abstract model to represent the goal we want to optimize, and address these problems using algorithmic and system technologies.

In this use case, we noticed that, while requiring query accuracy, the use mainly care about (1) Time till the first results, and (2) Time till the first a few results. This means to choose the data model and design the algorithms, we need to evaluate our solution using these two metrics, and decide which way to go.

Through this example of top relevant semantic search, we will learn how to apply models and techniques we learnt, to solve a real-world problem and improve our web-application's user experience. Indexing and caching are still powerful tools, we still care about system latency and throughput, we just need to adjust the definition of them for this special use case. We will also see new trade-offs made to meet these use cases.

5.2 Representation and algorithms

5.2.1 RDF data model

Web data can be represented using the Resource Description Framework (RDF) model. RDF is a standard model for representing data on the web in a structured and machine-readable format. It provides a way to describe resources, their properties, and the relationships between them.

RDF represents data as triples, which consist of subject-predicate-object statements. The subject represents the resource being described, the predicate represents the property or attribute of the resource, and the object represents the value or another resource that the property refers to. This triple format allows for flexible and extensible representation of data.

Web data can be converted into RDF by mapping the existing data elements to RDF entities and properties. For example, consider a web page that contains information about books, including titles, authors, and publication dates. This data can be represented in RDF using triples, such as:

```
<Book1> <Title> "Introduction to RDF". <Book1> <Author> "Xiaofeng Yang". <Book1>
<PublicationDate> "2023-2101-01".
```

In this example, `<Book1>` represents the resource (book), `<Title>`, `<Author>`, and `<PublicationDate>` represent the properties, and the values are represented as literals.

RDF provides a uniform way to integrate and link data from different sources on the web. It enables the creation of knowledge graphs and supports semantic querying, reasoning, and inference. By representing web data in RDF, it becomes easier to combine, share, and reason over heterogeneous data from various domains.

Representing web data using the RDF model allows for more structured and interconnected data on the web, facilitating data integration, semantic search, and knowledge discovery.

5.2.2 Search for top results

Here are a few approaches to searching for top results in RDF:

- SPARQL Queries. SPARQL (SPARQL Protocol and RDF Query Language) is a query language specifically designed for querying RDF data. SPARQL allows you to construct queries to retrieve specific information from RDF datasets. By formulating SPARQL queries with appropriate filters, sorting, and limit clauses, you can search for top results based on criteria such as property values, resource types, or other relevant parameters.

- Full-Text Search. If the RDF data includes textual content, we can employ full-text search techniques to search for top results, by extracting the relevant textual information from the RDF triples and use full-text search engines or libraries, such as Apache Lucene [16] or Elasticsearch, to perform efficient keyword-based searches. These search engines can return ranked results based on relevance scores or other ranking algorithms.

- Semantic Search. RDF allows for the representation of semantic relationships between resources. Semantic search techniques take into account the meaning and context of the data. This can involve utilizing ontologies or vocabularies to understand the relationships between resources and to infer connections. Semantic search can help retrieve top results by considering the semantics and relationships embedded within the RDF data.

Top-*k* keyword search algorithms for XML databases [17] combine semantic pruning based on document structure encoding with top-*k* join algorithms from relational databases. The main challenge lies in dealing with query semantics based on least common ancestors. RDF is a flexible and extensible way to represent information about World Wide Web resources.

Recent work on graph pattern matching in RDF databases has resulted in several different approaches to database design. However, as in the case of top-*k* query evaluation, it appears that more focus has been placed on scalability issues, such as replication, parallelization, and distribution of workloads [18, 19], as RDF datasets are often too large for a single machine. It is well known that SPARQL is descriptive enough to capture graph pattern matching queries (so-called basic graph patterns), and these queries are typically then decomposed into combinations of database primitive operations such as joins, unions, difference, etc.

5.2.3 Subgraph isomorphism problem

Searches on RDF requires efficient algorithms to search for subgraph on the data that is isomorphic to the graph representation of the query. Subgraph isomorphism is an NP-hard problem [20] in the size of subgraph, and state-of-art algorithms are not practical for large graphs. Lee [21] empirically compares the performance of several state-of-art subgraph isomorphism algorithms, including the Generic Subgraph Isomorphism Algorithm [22], Ullmann algorithm [23], VF2 [24], QuickSI [25], GADDI [26], and GraphQL [27]. They test on real-world datasets AIDS, NASA, Yeast and Human, covering a spectrum of relatively small graphs (with less than 1000 vertices). Modern social networks easily exceed that size by orders of magnitude, and the exact sub-graph isomorphism problem remains intractable for larger networks when label constraints and top-*k* are not fully exploited. Hence, to the best of our knowledge, none of these classic *precise* pattern matching algorithms could be used for our target application.

5.2.4 K-shortest simple paths

Similar to the subgraph isomorphism problem, the *k*-shortest paths problem is a natural and long-studied generalization of the shortest path problem, in which not one but several paths in increasing order of length are sought. The additional constraint of "simple" paths requires that the paths be without loops. This problem has been researched for directed graphs [28], undirected graphs [29], approximation algorithms [30], and algorithm engineering [31].

5.2.5 Anytime Top-k (Searches)

Finding the most important patterns plays a crucial role in web semantic search applications. What can we do to avoid running these NP-hard algorithms whose cost can easily grow exponentially as the query size? One way to go is to approximate the query results. However, remember that in our use case, we want to explore faster retrieval of top results while maintaining the complete correctness of query results. This lead to the path of employing new computational models that allow us to not compute all results.

Instead of paying the high cost of computing all results and then sorting them—we refer to that as *full enumeration*—the data analyst should be able to see the most relevant result "quickly", closely followed by the second-ranked, third-ranked, and so on. This way they can stop evaluation as soon as they are satisfied, or to pause and restart as often as desired. This means less computational resources and a faster response exploratory analysis. This property of being able to see the top-ranked results quickly will allow the analyst to try out many different ranking measures, and observe their effect on the output. This type of process is called term *any-k query*, for it returns the k top-ranked results without requiring the analyst to specify k in advance.

In traditional top-k algorithms, setting how many result to compute is a challenge: setting k too large wastes resources for computing irrelevant results and setting it too low requires restart with a larger k, again wasting resources. Some top-k algorithms can be extended to support incremental computation beyond the specified k.

Researchers focuses on any-k functionality for *conjunctive queries*, because they represent a sweet spot in terms of computational tractability and wide applicability [32]. The class of conjunctive queries includes subgraph homomorphism as well as natural joins with selections and projections, also known as SPJ queries. Joins are an essential building block of queries in relational databases. Exciting recent developments for joins between more than two relations and general join hyper-graphs have renewed interest in their efficient evaluation. We consider the same type of join queries as this recent work.

In a database, a join is an operation that combines rows from two or more tables based on a related column between them. It allows us to retrieve data from multiple tables simultaneously by establishing a connection or relationship between them.

Here is an example join in SQL:

```
SELECT column1, column2, ...
FROM table1.
INNER JOIN table2 ON table1.columnX = table2.columnY;
```

Joins are fundamental to relational databases and enable us to query and retrieve data that spans across multiple tables, providing a way to consolidate related information into a single result set. The most common type of join is an inner join, which returns only the rows that have matching values in both tables based on the specified join condition.

In order to evaluate the performance of an any-k algorithm, let us look at the following three measures:

1. Time-to-first (TTF): It returns the top-ranked result "quickly,"

2. Time-to-next (TTN): after returning a result, it returns the next result "quickly," and

3. Time-to-last (TTL): it takes "not significantly longer" to return all answers than the best-known algorithms for full query computation and ranking.

Once we have the models to reason about the algorithm performance, we can design solutions that meet user requirements and measure them. Here we are not going into the theoretical foundation of these optimizations. For interested readers who wants to know more about how these results are achieved, see full proofs and performance characterization in [32]. The researchers have proposed an algorithm that is proven to be optimal in the worst case of RDF data input. This means by implementing the algorithm in the RDF database, the application engineers are guaranteed that Time-to-first (TTF), Time-to-next (TTN), Time-to-last (TTL) are significantly lower or comparable to all other algorithms.

6. Summary

In this special application, we went through the use case of interactive semantic searches on the web, and discussed the software engineering steps happened before the design and implementation phase, i.e. user study, key use case analysis, modeling the use case into an abstract problem.

Real-world applications come in all shapes and flavors, but we hope that the reader have gained some insight in how to approach a problem by disassembling it into smaller pieces of (1) use case study, (2) product requirement design, (3) system design, evaluation, then (4) deployment and maintenance. Once we have a formulation of the engineering problem at hand, there are lots of tools and technologies to choose from to solve well formed sub-problem in each step. With the user requirements in mind, we can evaluate all these tools using methodological perspectives, and find solutions that work the best for our use case.

Enjoy the journey!

Author details

Xiaofeng Yang
Google, Mountain View, USA

*Address all correspondence to: xiaofeng@ccs.neu.edu

References

[1] Emerson S, L, Darnovsky M, Bowman J. The Practical SQL Handbook: Using Structured Query Language. Boston, USA: Addison-Wesley Longman Publishing Co., Inc; 1989

[2] George S. Nosql–not only sql. International Journal of Enterprise Computing and Business Systems. 2013; **2**(2):2-5

[3] Maule A, Emmerich W, David S. Rosenblum. Impact analysis of database schema changes. In: Proceedings of the 30th International Conference on Software Engineering. New York, United States: Association for Computing Machinery (ACM); 2008. pp. 451-460

[4] Stonebraker M. Sql databases v. nosql databases. Communications of the ACM. 2010;**53**(4):10-11

[5] Han J, Haihong E, Le G, Jian D. Survey on nosql database. In: 2011 6th International Conference on Pervasive Computing and Applications. Port Elizabeth, South Africa: IEEE; 2011. pp. 363-366

[6] Lakshman A, Malik P. Cassandra: A decentralized structured storage system. ACM SIGOPS Operating Systems Review. 2010;**44**(2):35-40

[7] Banker K, Garrett D, Bakkum P, Verch S. Mongo DB in Action: Covers Mongo DB Version 3.0. New York, United States: Simon and Schuster; 2016

[8] Chang F, Dean J, Ghemawat S, Hsieh WC, Wallach DA, Burrows M, et al. Bigtable: A distributed storage system for structured data. ACM Transactions on Computer Systems (TOCS). 2008;**26**(2):1-26

[9] Miller JJ. Graph database applications and concepts with neo4j. In: Proceedings of the Southern Association for Information Systems Conference. Vol. 2324. Atlanta, GA, USA: Association for Information Science and Technology; 2013

[10] Bach M, Werner A. Standardization of nosql database languages. In beyond databases, architectures, and structures. In: 10th International Conference, BDAS 2014, Ustron, Poland, May 27–30, 2014. Proceedings 10. Ustron, Poland: Springer; 2014. pp. 50-60

[11] McGrath G, Brenner PR. Serverless computing: Design, implementation, and performance. In: 2017 IEEE 37th International Conference on Distributed Computing Systems Workshops (ICDCSW). Atlanta, GA, USA: IEEE; 2017. pp. 405-410

[12] Varia J, Mathew S, et al. Overview of amazon web services. Amazon Web Services. 2014;**105**:5-11

[13] Baron J, Kotecha S. Storage options in the aws cloud. In: Amazon web Services. Washington DC: Tech. Rep; 2013

[14] Astrahan MM, Blasgen MW, Chamberlin DD, Eswaran KP, Gray JN, Griffiths PP, et al. System r: Relational approach to database management. ACM Transactions on Database Systems (TODS). 1976;**1**(2):97-137

[15] Edgar Frank Codd. A relational model of data for large shared data banks. Communications of the ACM. 1983;**26**(1):64-69

[16] Białecki A, Muir R, Ingersoll G, Imagination L. Apache lucene 4. In SIGIR 2012 Workshop on Open Source Information Retrieval. Vol. 17.

New York, United States: Association for Computing Machinery (ACM); 2012

[17] Chen LJ, Papakonstantinou Y. Supporting top-k keyword search in xml databases. In: Proc. ICDE. Piscataway, NJ, USA: IEEE Institute of Electrical and Electronics Engineers; 2010. pp. 689-700

[18] Huang J, Abadi DJ, Ren K. Scalable SPARQL querying of large RDF graphs. PVLDB. 2011;**4**(11):1123-1134

[19] Hose K, Schenkel R. WARP: Workload-aware replication and partitioning for RDF. In: Proc. ICDE Workshops. Piscataway, NJ, USA: IEEE Institute of Electrical and Electronics Engineers; 2013. pp. 1-6

[20] Lubiw A. Some NP-complete problems similar to graph isomorphism. SIAM Journal on Computing. 1981;**10**(1): 11-21

[21] Lee J, Han W-S, Kasperovics R, Lee J-H. An in-depth comparison of subgraph isomorphism algorithms in graph databases. PVLDB. 2012;**6**(2): 133-144

[22] Cordella LP, Foggia P, Sansone C, Vento M. A (sub)graph isomorphism algorithm for matching large graphs. IEEE Transactions on Pattern Analysis and Machine Intelligence. 2004;**26**(10): 1367-1372

[23] Ullmann JR. An algorithm for subgraph isomorphism. JACM. 1976; **23**(1):31-42

[24] Cordella LP, Foggia P, Sansone C, Vento M. An improved algorithm for matching large graphs. In: Proc. IAPR-TC15 Workshop on Graph-Based Representations in Pattern Recognition. Piscataway, NJ, USA: IEEE Institute of Electrical and Electronics Engineers; 2001. pp. 149-159

[25] Shang H, Zhang Y, Lin X, Yu JX. Taming verification hardness: An efficient algorithm for testing subgraph isomorphism. PVLDB. 2008;**1**(1): 364-375

[26] Zhang S, Li S, Yang J. Gaddi: Distance index based subgraph matching in biological networks. In: Proc. EDBT. New York, United States: Association for Computing Machinery (ACM); 2009. pp. 192-203

[27] He H, Singh AK. Graphs-at-a-time: Query language and access methods for graph databases. In: Proc. ACM SIGMOD. New York, United States: Association for Computing Machinery (ACM); 2008. pp. 405-418

[28] Yen JY. Finding the k shortest loopless paths in a network. Management Science. 1971;**17**(11):712-716

[29] Katoh N, Ibaraki T, Mine H. An efficient algorithm for k shortest simple paths. Networks. 1982;**12**(4):411-427

[30] Roditty L. On the K-simple shortest paths problem in weighted directed graphs. In: Proc. SODA. New York, United States: Association for Computing Machinery (ACM); 2007. pp. 920-928

[31] Hershberger J, Maxel M, Suri S. Finding the k shortest simple paths: A new algorithm and its implementation. ACM Transactions on Algorithms. 2007; **3**(4):45-es

[32] Tziavelis N, Ajwani D, Gatterbauer W, Riedewald M, Yang X. Optimal algorithms for ranked enumeration of answers to full conjunctive queries. In: Proceedings of the VLDB Endowment. International Conference on Very Large Data Bases. Vol. 13. New York, United States: NIH Public Access; 2020. p. 1582

Chapter 10

Paper Recommender System Using Big Data Tools

Nasrin Jokar, Mehra Esfandiari, Shima Aghamirzadeh and Hossein Hatami

Abstract

To face the problem of information overload, digital libraries, like other businesses, have used recommender systems and try to personalize recommendations to users by using the textual information of papers. This textual information includes title, abstract, keywords, publisher, author and other similar items. Since the volume of papers is increasing day by day and recommender systems do not have the ability to cover this huge volume to process papers according to the user's tastes, that is why we need to use our papers to cover and process this volume quickly. We have big data tools, which will offer relevant recommendations by running parallel processing. In this chapter, the researches and researches of researchers in the field of recommender systems/aware of the text of scientific papers and recommender systems have been discussed.

Keywords: paper recommender system, personalization, textual information, big data, extracting user interests

1. Introduction

Scientists usually rely on sharing their research results to obtain peer, research and financial opportunities. The most common approach to share knowledge is publication. In the past decades, academic publications have had significant changes since electronic format replaced with printing format. The advent of Internet makes the possibility for electronic publishers to publish more research in the area. Hence, the distribution of publications has generally become simpler and the widespread publication of the Web has increased at higher levels of publications. Researchers often share their research through the web, journal, papers, conference and presentations currently. Researchers determined three types of websites for distribution and publication of scientific papers: personal, institutional and social. Personal and institutional websites are created, respectively, by scholars and academic institutions to provide academic information. Social web sites allow researchers to automatically submit papers and research results; for example: LinkedIn, Academia.edu, Research Gate, Google Scholar [1].

Many researchers publish scientific papers without time and place restrictions. Users find themselves in front of hundreds of thousands of scientific-research papers and choose a sample of them according to their tastes and needs. But when users are

IntechOpen

encountered with a huge amount of information, the selection problem arises and searching among this space takes a lot of time. Recommender systems are effective options to direct and guide the user, among a huge amount of possible choices to obtain the useful and favorite option. Recommender systems for research papers have become increasingly popular. In the last 14 years, more than 170 research papers, patents, web pages, etc. have been published in this area. Recommender systems are useful applications for research papers, for example, they help researchers who are pursuing their research field [2].

Nowadays, there are many approaches for scientific recommendation on such papers [3], which most of them are based on two content-based (CB) algorithms and Collaborative refinement, but there are two other methods, knowledge-based (KB) algorithm and combination method that can be a combination of three other methods. The main idea of the CB methods in the paper recommendation systems, the user's interest and priority of study that has previously studied, is specified by the content of scientific research. In this way, although the focus is on finding similarities between items or scientific papers, the collaborative refinement methods perform as extraction and using similar user's feedback without analyzing the content and focus on similarity between users. These approaches have been introduced and evaluated in more than 170 research papers, as well as patents, lectures and blogs [3].

Papers recommendation systems are content based and reference-based that return results by keywords. Contextual-based recommendations are usually carried out using specific keywords provided to search engines such as CiteSeer and google scholar [4]. The referral-based recommendations is aimed to help authors in selecting most relevant papers for referring to the potential number of resources, that referencing relevance depends on the content of the paper; For referral, the contextual next to the bracket ([]) is used, in such a way that a certain number of words are used before and after referrals, few used sources demonstrated by the recommendation engine in the predefined [5]. In referral recommendation, the concept of referrals and the content of papers (documents), there is a completely different nature, and the translation model can be a bridge between these gaps.

The Web-based recommendation systems are highly interested because personalized services are important factors for shopping malls and Internet service providers. Researchers and academics save their research information and others' on the Internet and share them among each other and eventually organize them. In the references [6, 7], a personalized recommendation system has been introduced that increases the accuracy of research papers, and it can be a specific, feasible and applicable domain for developing recommendation system.

With regard to the fact that in a system, users usually share different domains of papers that they have specific preferences for specific domains, in reference [8], it is assumed that the user interests can be extracted from their behavior on the site, which is calculated (when they visit pages). By collecting and analyzing user's behavior, they have approached user's interests and created a personalized recommendation model that selects nominated papers for recommendation. Their experiments demonstrated that the user-behavioral recommendation model improves the accuracy of the recommendation. If users have just registered or they are not active or even they have little behavioral information, the system can't receive knowledge about their priorities. Also, for implementation of user-centered personal recommendation, the system should track the behavior of the user to collect information that includes user's searches, scope of work, published papers as well as friends or people associated with them in this field.

Context-aware recommender systems are systems that try to provide each user with suggestions based on performance, personal tastes, and user behaviors and depending on the fields in which they are used. To help in the decision making process. Background information extracted for digital libraries (DLS) are classified into three main groups of users, documents and environment [9]. Users' textual information includes user profile, user type, purpose, activity/work, knowledge/skills, social networks and home page information, logs, information behavior and level of information need. For example, it means the type of user, student, faculty member, researcher or researcher, or the level of information need means whether the user needs it urgently or not. Each paper has special features that are different from other papers. These features can be considered as the background information of the paper. Background information includes bibliographic information, references between papers and popularity of papers. Contextual information also includes the location of items in the library or geographical location, time and type of service.

The bibliographic information of the papers includes title, number or abstract International Standard Serial Number (ISSN), keywords, author name, publisher, publication time, paper classification, paper main text, paper format, language, URL and paper status in terms of rank [9]. In the researches [2, 10–12], the information of the title, abstract, keywords and the main text of the papers have been used to find the most similarity of the user's query to the data set, the researches [7, 13–15]. It has used the text of the paper for recommendation Research [7, 8] has used keywords for recommendation. Research [16] has used the title of the papers to create a profile of the papers to categorize them, in addition to using the title, the author's name is also used for categorization, but since in each research the researcher has access to all the papers published in does not have the world, this feature is used in handfuls, which Sun et al. [13] have used this method.

Reference background information between papers includes papers cited by the paper or papers, which researches [4, 14, 17, 18] have used the reference of papers in the method presented in their papers. The background information of paper popularity includes the most popular papers in the week/month/year, which only research [5] used this feature. In the existing dataset, such information is very rare and cannot be manipulated much on this feature. Since the data set used is limited to a few journals, only some bibliographic background information of the paper has been used.

Text-based and citation-based paper recommender systems return results. Text-based recommendations are often made of specific keywords provided to search engines such as Citeseer, IEEE, and Google Scholar [4]. In citation-based recommendations, the citation recommendations is aimed to help authors select the most relevant papers for citation from a potential number of references, where the citation relevance depends on the paper content. The placeholder text shown in the form of brackets "[ ]" is used as a context for the reference, in that a certain number of words are used before and after the reference, which are used in the pre-definition of a small number of sources. Which is used by the recommender engine [5].

Since the volume of papers is increasing day by day and recommender systems do not have the ability to cover this huge volume to process papers according to the user's tastes, that is why we need to use our papers to cover and process this volume quickly. We have big data tools, which will offer relevant recommendations by implementing parallel processing (map-reduce programming).

2. A review on paper recommender systems

In this section, we have examined some of the researches and researches of researchers in the field of recommender systems that are aware of the text of scientific papers and recommender systems, also the researches that are aware of the text have also been used. These researches include a systematic review on text-aware recommender systems, recommending papers for reference to draft papers, personalized recommendations, extracting user interests and behavior and analyzing them for personalized recommendations and extracting the domain of papers, as well as the recommender system of the papers is divided into the big data platform, that show in **Table 1**,which will be explained.

2.1 Text-aware recommender systems

Recommender system is necessary as a suitable tool to facilitate and accelerate dynamic information processing due to the significant increase of data in DLS. In recent years, recommender systems use the description of users' situational information such as location, time, work, etc. in order to make more relevant and personalized recommendations, and to predict accurate recommendations to users in a specific domain such as DLS, It is essential to understand and exploit the relevant conditions of the users, which leads to the creation of an intelligent recommendation.

Reference [10] is a systematic review on research text-aware recommender systems, the purpose of this study is to conduct a literature review on recommender systems for academic DLS for the following:

a. Identification of the textual information adopted to build the recommender in the academic DLS.

b. To identify the methods that have been used to adopt textual information for making recommenders in academic DLS.

Method	Used by	Description
Text-aware recommender systems	[10]	The background information (context-aware) of articles and users is used to make recommendations
Cite to draft papers	[4, 10, 11, 14, 19, 20]	Citation recommendation systems generate a list of relevant articles to cite in a particular text.
Personalized recommendations	[6, 7, 16, 21, 22]	The personalized recommendation system based on the query and the text of the articles offers recommendations that increase the accuracy of finding research articles
Extracting user interests and behavior	[8, 22, 23]	This recommendation system extracts the user's interests based on the user's behavior and suggests recommendations according to the user's interests
Extracting the domain of papers	[10, 13, 24, 25, 28–31]	Extracting the domain of articles and creating profile of articles for recommendation
Big data platform	[30, 31]	They are implemented using big data system platforms, which are suitable for systems with huge data

Table 1.
Overview of recommender systems surveyed.

c. How has the relevance of textual information in recommender systems for the academic domain previously used by researchers been discovered and understood.

The main contribution of this research is to review past studies in order to discover textual information that is effective on recommendations in the domain of academic DLS. They reviewed 82 papers published from 2001 to 2013 to identify background information and methods used to develop scientific recommendations. The authors of the paper have conducted this review in three phases to identify textual information: the first phase: planning review, the second phase: conducting the review, and the third phase: review report. Finally, the results of their investigation showed that the contextual or textual information extracted for recommender systems in academic DLS is classified into three main groups, which include the user, documents, and contextual information environment. The background information of the user indicates the information of the user's current status, such as user profile, types of users, purpose, activity/work, prior knowledge/skills, social networks and homepage information, logs, information behavior (searching, reading, etc.) and the level of information requirement (necessary and unnecessary). Each document has special characteristics that are different from other documents, such as bibliographic information (title, abstract, ISSN, etc.), references, popularity (number of visits or downloads of a paper per month or year), which these characteristics are known as information. Background documents are adopted. Contextual information environment provides a set of information for users' official status such as location, time, service type, especially when users' status changes dynamically and frequently.

To answer question **b**, they divided the many methods that have been successfully applied to produce recommendations into four categories: common filter (CF), CB, KB, and other methods (data mining, cluster analysis, etc.). To answer question **c** through five methods of contextual communication in the recommendation, which include the following: researchers have referred to the past information used in this contextual information, as well as other researchers' context definitions.

2.2 Cite to draft papers

When a writer is writing or preparing to write a research paper, he always has the appropriate sources in mind, however, there are likely to be sources that the writer has missed that should be cited. In this way, a good reference recommender system will not only improve the paper, but in general, it will improve the efficiency and quality of literature search, which in this section we will examine some examples of this type of recommender systems.

In reference [26], in order to recommend papers to refer to the draft paper, it first creates a set of documents for ranking in three stages, which includes the following steps: The first stage selects 100 papers that are most similar to the draft paper. It chooses R as the set. In the second step, it adds all the papers referenced in R to the collection (finds the sources papers and adds them to the collection). In the third step, it adds all the papers that are referenced by some papers in R to the collection. Finally, it uses 0 existing documents in the R collection with features such as publication year, similar text, same author, co-referencing pairs, Katz distance measurement graph and reference count (number) for ranking. According to the test results, they introduced the following features as important features in the final ranking for better system performance, which are: author, number of references, publication date, title text and Katz measurement function.

Research [10] has provided a text-aware referencing recommendation for a draft paper considering the inability of graph-based link prediction technique and natural language processing analysis. He presented the problem in such a way that d represents the document, D represents the collection of documents, a context C which is a set of words around the bracket that is to be referenced, a global context which includes the title and abstract d, and a local context which is the surrounding text. The reference is in d. Also, if document d1 refers to document d2, the local context of this reference is named out-link, and considering that document d1, document d2 is called the context of in-link. According to draft paper d without bibliography a national recommendation is a ranked list of references in set d that are recommended as candidates for bibliography d. Given out-link local context, c* given d, a local recommendation is a ranked list of referrals in the set D recommended as candidates for the holder " [] " associated with c*.

Their framework-based recommendation system [10] can take two types of input, a manual query d1 or just an out-link local context for c*, which they use to extract its global text (e.g. abstract and title) and All out-link local contexts process it. Another entry for each reference holder associated with an out-link local context is provided for the bibliography, ranking the references by their relevance and returning the top k papers.

Research authors [10] use two methods to retrieve papers from their candidate collection (CiteSeer library papers).

Text-aware methods, in which the local context is not considered in this method, and for a manual query d1, documents can be retrieved using the following methods:

1. N top references with abstracts and titles that are most similar to d1, which is called GN.

2. Documents that authors share with d1, which is called the Author method.

3. Papers referenced by documents previously produced in the candidate collection by some other methods (for example, GN, Author), which is called CitHop.

4. Documents written by authors whose papers are currently in the candidate collection and are generated by some other method called AuthHop.

The above methods are too slow because they are based on the method based on the text content of the documents and may cover papers that are not relevant and appropriate (many ideas are not in the abstract).

Context-aware methods, which help to improve the coverage for recommendations by considering the local context in the manual query. For a d1 for each local context c* they introduced two methods:

Top N documents that are most similar to c* in terms of in-link, which is called LN.

Documents containing N out-link fields that are most similar to c* (this method refers to documents retrieved by LN), this method is called LCN.

The above 6 methods can be combined. They used Gleason's non-parametric probabilistic model to measure the association between a draft document and a referral candidate and to rank it.

Given the manuscript document d1, the authors used formula 1 to rank the documents for recommendation in the c* context related holder. If the reference context c* is given without d1, then we use the formula 1.

$$\mathrm{Sim}(\mathrm{c}*, \mathrm{d2}) = \mathrm{Trace}(\mathrm{Td2\ c} * \mathrm{c} * \mathrm{T}) = \frac{1}{k2} \tag{1}$$

When an author writes a paper, he uses words and methods that he forgets which author he is from. If the recommender system can provide some candidate papers containing the required information about the use cases, writing the paper becomes easy. In the research [14], the referral recommendation problem has been improved with the translation model. Basically, the translation model is used in one language to another language, and it is also used in document retrieval that displays heterogeneous content in search (for example, cross-language retrieval). In the problem of referral recommendation, the concept of referral and the content of papers (documents) are actually completely different in nature and the translation model can be a bridge between this gaps. This method actually works better than the state of art method and has better performance.

Many researchers have studied the application of CF algorithms in citation recommendation for papers. However, CF algorithms have limitations such as sparse data and scalability, and in referral recommendation tasks, referral graphs tend to be noisy and sparse, and potentially due to referral errors, referrals are lost or have space limitations. To deal with these problems, the authors of reference [4] have used Singular Value Decomposition (SVD) method to build a reliable referral recommendation system and address the limitations of memory-based CF algorithm. SVD is a popular method for identifying latent semantic factors, where association patterns in data can be easily identified and compared to the original space.

People are faced with the problem of information overload on a daily basis. As people publish more information on the World Wide Web, this problem is getting worse day by day, making it difficult to find the information they need. By recommending items to users based on previously stated user preferences, recommender systems guide and control the complexity of users' information space. Research [18] integrated the CF algorithm in the domain of research papers that works with a dataset as a ranking matrix, where the rows of the ranking matrix represent users and the columns of the matrix represent items. In order to integrate the CF algorithm in the domain of research papers, the reference network on the CF score matrix is needed. There are different ways to create a CF score matrix from the reference network between research papers.

One solution would be the references as items in the matrix while the users would be the rates of the papers, which approach is not used in the reference network due to the setup problem. An alternative approach is to consider the authors of papers as users and the maintainers of references as items. In this author's matrix, the references used by that author are given points in the matrix. The problem with this method is that an author may have written several papers in different fields during his career, which may provide an incorrect recommendation system [18].

The recommendation algorithm proposed in the paper [18] uses a referral basket as input to the given system and the system returns a ranked list of referrals as output. In this experiment, co-referencing algorithms such as co-referencing matching, CF user-item algorithm, CF item-item algorithm and simple Bayesian classification were used against non-CF algorithms such as local reference search graph and keyword search.

The test has been evaluated offline and online. In the offline evaluation of the recommender system based on the Bayesian algorithm, it has the best result in terms of ranking, and considering the coverage of the papers, the user-user and user-item algorithms have performed the best. In the online test, which are categorized according to related and unrelated papers, Bayesian methods, local reference search graph, and keyword search have performed better, but in the newness of papers, user-user algorithms and user-item algorithms have performed better. Finally, CF algorithms are able to make recommendations for each paper and are suitable for reading, while basic methods are suitable for finding related works.

Citations are very important for the credibility of scholarly papers, as well as proper references to support claims in your work. However, as the number of research publications grows, researchers may find it difficult to find suitable and necessary references, so using a reference recommendation engine can recommend authors writing a paper a complete list of suitable and relevant references. Many reference recommendation models are divided into two categories: global recommendation, which recommends a list of references for a draft paper, and local recommendation, which recommends references for a specific topic in the paper. Reference [11] focused on local recommendations or context-based recommendations. A citation context C is defined as a sequence of words that appear around a specific citation, usually a citation context containing words that describe or summarize the referenced papers. Directly, the semantics of referenced documents should be close to the reference context. The authors of the paper [11] use a neural network model to estimate the probability of referring to the paper according to the reference context. This method ensures that keywords used to reference similar documents have a high semantic similarity. To evaluate this model, a general test was performed on the CiteSeer dataset, which performed better than the State of Art method.

2.3 Personalized recommendation

Interest in web-based recommender systems is high because personalized service is an important factor for shopping malls and Internet service providers. Researchers and academics store and share their research information and others on the Internet, and finally they organize. In the research [6, 7], a personalization recommendation system was presented that increased the accuracy of finding research papers, also this system can be feasible and applicable for the development of a recommender system of a specific domain.

The proposed system architecture [6, 7] consists of user interface, crawler, transform, extractor, filter and keyword finder, management profile, user profile and database. The user interface interacts with users who create input and output topics for each topic for the crawler. The crawler of papers from Google Scholar stores the URL of each paper in one place and converts them all to a specific format. The extractor extracts the paper title, keywords, abstract and main body of each research paper in text form. If a paper does not have keywords, we extract keywords from the title of the paper. Filtering process, suitable papers for users among the stored papers according to the topic selected by the user through the user's profile, which includes the user's personal preferences in the form of saved XML forms, the filtering process is done.

The suggested algorithm for extracting keywords is that first unnecessary characters such as or, of, is, etc. are removed, then the keywords in the title and introduction, keywords specified for the paper by the author and keywords on the first page

Papers are extracted. If there is no keyword, a few words can be selected from the title and the formula 2 checks the degree of reflection between the title and the keywords:

$$\text{Reflex} = \frac{\#\text{Keyword Title}}{\#\text{Title Term}} \quad (2)$$

A user profile is collected with each user click on a research paper. Each time a paper is selected, the frequency of domain, topic, and profile keywords is incremented, and then the rate of each occurrence is recalculated, reflecting the user's profile. To calculate the similarity, cosine similarity is used, which is generally used to solve the cold problem. To evaluate the test, SAT, which means the ratio of the total number of correct and incorrect research papers to the total number of research papers stored for a certain subject, and ACC, which means the ratio of correct research papers to the number of recommended research papers, have been used, which have had significant results.

In research [16], for each user, it provides a small set of categories as the context for each query sent by the user, which is based on their search history. Their method is as follows: (1) A strategy is used to collect the user's search history, (2) A user profile is created based on the search history of a public profile, (3) Appropriate categorization for each user query based on the user's profile. And public is chosen.

To personalize the search, the user's search history must first be created, which is a tree of search records, the root of the search history is a query, and each query has one or more related categories. Then, a user profile is created that shows the user's interests, and finally, a matrix is created for the user's history and profile, and a general profile is created from these two matrices. . In addition to using the three mentioned matrices, general knowledge that is applicable to all users has also been used. Personalization is done by mapping a user search to a set of categories, reflecting user intent based on user profile and general profile. The maps are as follows:

1. The similarity between the user's query and the categories representing the user's interests is calculated.

2. The similarity of categories is ranked in descending order.

3. Finally, it shows the top three categories together with a button that shows the next three categories to the user.

In the study [21], the combination method in the individual prediction stage and the aggregation method based on the ER rule in the group aggregation stage are proposed to recommend the improved GPRAH_ER group article. The framework of the researchers' proposed approach consists of three steps: The first step: data acquisition, where the original data is obtained by the web crawler as in Jokar et al.'s [30] paper, which is used to retrieve all researchers, articles, and groups in SSNs. Records of collecting articles, researcher group relationships, content of articles and other information are collected together. Then remove data noise, segmenting the words, and removing the stopped words are performed for the obtained data. In the second stage, the individual prediction method is performed based on the probabilistic matrix factorization (PMF), (that it is a powerful method for modeling data associated with pairwise relationships) method, one of the common model-based CF methods, which includes secondary information from articles and group information. The decomposition process is performed based on the researcher-article ranking matrix $R_{M \times N}$ together with the researcher-group matrix

$G_{M \times L}$ to obtain the researcher's hidden feature matrix $U \in R^{M \times K}$, the article's hidden feature matrix $V \in R^{K \times N}$. The researcher-article ranking matrix $R_{M \times N}$ is decomposed into the researcher latent feature matrix U and the article latent feature matrix V. U_i and V_j refer to the latent feature vector of the ith researcher and the latent feature vector of the jth article, respectively. To ensure that articles with similar content can have similar feature vectors when doing matrix factorization, the content similarity between articles is calculated based on the content-based filtering (CBF) (CBF algorithms recommend suitable items to users based on the descriptions of items and user preferences) method and then included in the PMF method. Finally, in the third step, the group aggregation method is used. That the group aggregation method is built for each group to obtain the predicted final ranking of the groups through the evidential reasoning (ER) approach. For each group, the final predicted rankings on the articles are sorted in descending order, and then the top-ranked articles are recommended to the groups. Thus, ER is a general approach for multi-criteria decision analysis that can be considered as a probabilistic approach by using an integrated structure to model different types of uncertainty and from all the data generated by users. Full use. Researcher-article $\tilde{R}_{M \times N} = \{r_{ij}\}$ is obtained by the combined method. Through the $\tilde{R}_{M \times N}$ matrix, the prediction values of the group members in the articles (i.e. r_{ij}) can be shown. Then, the aggregation process which merges the final recommendation list of each group by the prediction values of all members by the ER rule is performed. For each UI in group G_k, his prediction values across all articles can be taken as evidence for the "truth" prediction that that article matches the group's taste.

2.4 Extracting user interests and behavior

With regard to the fact that in a system, users often share papers from different domains given the specific priorities for specific domains. In the research [8] it is assumed that the user's interests are extracted from their behavior on the site calculated automatically (when they visit the pages). They approached user interests by collecting and analyzing user behavior and developed a personalized recommendation model that selects candidate papers for recommendation. Their experiments showed that the recommendation model based on user behavior increases the accuracy of the recommendation. The system has little behavioral information about newly registered or inactive users. It cannot get knowledge about their preferences, so they optimize their proposed model.

To implement user-centered personal recommendation, they need to follow the user's behavior to collect information and discover the feature that can reflect the user's feature, the preferences with high influence in the model should be selected. They [8] select user preferences from the site structure and their existing data records. Active user operations include publishing a paper, marking a paper as interest, rating a paper, creating a comment, and tagging a paper. They [8] extract the title, abstract, keywords of papers of interest to each user and configure them in one file. Also, they have defined the recommendation task as a triple relation D_i, D_x, U, where U refers to the user, D_x refers to the collection of papers that the user is viewing, and D_i refers to the collection of papers for recommend to the user.

In this work [23], a user intention model based on deep sequential topic analysis is proposed. This model predicts user intent based on the topic of interest. The proposed approach is classified into two steps: (1) user preference learning and (2) user intent prediction. In the first step, the learning method implicitly explains the user's preferences. The proposed method gathers the user's preferences implicitly by considering the user's click data. In the research article recommendation system, clicked data points to a set of articles. A paper has several properties. The proposed method considers the latent features of articles that are "topic" as users' preference. In other words, the proposed approach tries to capture the user's preference in terms of topics of interest. In other words, the proposed approach tries to capture the user's preference in terms of topics of interest. In order to extract the topic of the articles, a hybrid model is proposed. The subject model is a combination of Latent Dirichlet allocation (LDA) method and Word2Vec method. To calculate the probability distribution of the words of an article among several predefined topics, LDA, a topic modeling technique, is considered. The measure of word-to-word correlation is necessary to justify the true identification of essay topics. In this context, a Word2Vec word embedding technique is considered. Word2Vec helps to group semantically and syntactically similar words in a specific topic. This work proposes a method how to combine LDA and Word2Vec to obtain the full word-topic distribution. In the second stage, the user profile is divided into two types: (1) static and (2) dynamic. A static profile contains such user information (such as age, gender) that does not need to be changed. Generally, profile information is provided by the user himself. In contrast, the dynamic profile is automatically generated by the recommendation system, and the features it contains undergo changes over time. Since, the proposed approach captures a user's interest topic that may change in different context or time, it is dynamic. It will be updated accordingly, and its size and variety will also increase. In this regard, conventional user profile modeling approaches are used to obtain all historical sequences (long-term and short-term) of user-item interaction. The proposed approach considers a deep sequential topic analysis technique to predict a user's future favorite topic from his past preferred topic sequence. Specifically, a variation of Recurrent Neural Network (RNN) which is Long Short-Term Memory (LSTM) is used to combine long-term and short-term topic interest to predict future topic interest. This model considers the time context, i.e. the time difference between the clicks of two consecutive articles viewed by the user. In addition, a few external attributes such as "likes" (whether the user likes the article or not), "session number" are considered.

2.5 Extracting the domain of papers

Text analysis is a process of analyzing unstructured textual data with the aim of extracting meaningful information. This meaningful information is used to identify the domain of each paper, which domain refers to a specific branch of scientific knowledge or scientific field. For example, the scientific fields of computer science have several domains such as data mining, network, operating system, etc., and the domain of data mining, in turn, has several sub-domains such as pattern recognition, machine learning, statistics, etc. Any given paper basically has a set of words, when the researcher reads the paper, he may be able to discover its domain and subdomain, but the ability to understand this subject can be in the researcher's previous knowledge of the domain and subdomain.

In reference [27], disambiguation of prepositions using supervised learning method is used to extract the scope of each paper. In this way, they have described a method for extracting meaning from the title, keywords and abstract of research papers in the field of computers.

To identify the scope of each paper, transitive methods such as the use of words, stop words, sentences, clauses, phrases, m grams, prepositions, prepositions with a sense of intention, favorite phrase, derivative, word scope, disambiguation of prepositions and etc. there are The definition of some of these items is as follows:

Sentence: It is a sequence of words that are complete by themselves and contain a subject and proposition and convey a statement, and it consists of a main clause and has one or more clauses.
Clause: It is a unit of command organization that includes entity and proposition.
Preposition: A ruling word that usually comes before a noun or pronoun and expresses the relationship with another word or element in the clause.
Preposition with sense of intention: It is a preposition that follows a phrase that indicates a specific purpose or action.
Favorite phrase: phrases that end after a preposition with a sense of intention and before the next preposition in the sentence or with the end of the sentence.
Disambiguation of prepositions: Semantics is a branch of linguistics that deals with the meaning of words and phrases in a particular context.

In the proposed method [27], the title of a paper, which can be a sentence or a phrase, is used, and the title is scanned to find prepositions with a sense of intent, and interesting phrases that follow a preposition are extracted. In the next step, interesting phrases that have one or more words in common with the keywords are saved as practice words. Each derivative is then labeled and classified as seed or non-domain, which forms the dataset for training data as a gram list. The test data is also classified using the training data. Text mining is a growing field that involves automatic knowledge extraction from natural language. Natural language processing or NLP is a set of essential techniques for extracting text, which disambiguation of prepositional "concept" is an NLP method that researchers extract the important "concept" conveyed by prepositions in the text.

In the study [28], the important concepts of prepositions were extracted from the text with some interesting expressions, and these interesting expressions were then classified using a simple Bayesian classification. The domain of a scientific paper is the main topic or topic raised in it, which is a branch of a scientific discipline, for example, the field of computer is domains such as network, parallel computing, theory of computation, data mining, etc. Each domain can in turn have other subdomains, in which case the second domain is considered, for example, the second domain for the data mining domain includes NLP, pattern recognition and . . . Is. The method used in this research [28] is to extract the root or context of each paper in the form of its domain, and interesting phrases based on their place in the vicinity of certain prepositions are deduced using the results of preposition disambiguation.

The authors of the paper [28] have used interesting phrases for their research, they have compiled a complete list of prepositions after reviewing several English textbooks. Each preposition in this set is defined as pi, with all other prepositions defined as ts:

$$\text{Pi} = [\text{for, to, towards, ...}]$$

Complement C is an expression that is derived based on the permutation of pi and p0 in a clause, and E is defined as the end of a clause:

$$\text{pi C p0, p0 C pi, pi C E}$$

Authors of papers [28] have used integration and keywords to extract interesting phrases, because the titles of papers tend to be unique and are good proxies for domain recognition, and on the other hand, phrases or keywords are usually more widely used, which Used for tagging to identify the domain. Then the common interesting phrases in the title and keywords are retained as derivatives for retrieval. For classification, a small amount of input data is labeled and each derivative is classified as "domain" or "not domain" using a Bayesian algorithm.

In the research [29], the title and abstract were used to summarize the paper, which were presented based on the identification of frequent phrases in the titles of research papers to discover the trend of research topics in a field. In this method, they first extract phrases by removing punctuation marks and stop words. These phrases may be repeated in many similar lines, which should be extracted from these phrases to determine how many times each sub-phrase is repeated in the title and abstract. The proposed method has been implemented with Java and R languages for various conferences and journals, which have had significant results.

The authors of reference [13] proposed a personalized recommendation system based on the interest survey model, which effectively examines the characteristics of newly published papers and selects the appropriate group for them. Personal recommendation technology is mainly divided into three categories, including rule-based filtering, content-based filtering, and collaborative filtering.

Rule-based filtering requires users to provide information about their interests, whose interest model is created and maintained by them. The advantages of this method reflect the accuracy of the user's interest, but due to the availability and scalability of such systems, they may perform poorly because the users take responsibility for the modeling. Building a model for content-based filtering is such that the system automatically collects user history information, the problem of this method is to build an ontology domain, because it requires a large amount of training text for clustering. The joint filter is usually based on KNN, which is very popular, but it currently has challenges such as sparse data and scalability, and the growth of the number of users and cases leads to computational complexity and also suffers from the cold problem. Due to the problems that the above methods had, the authors have developed a personal recommendation method that uses classification algorithms in machine learning (the Bayesian hierarchy model for content-based recommendation).

The article [25] was done in three stages, in the first stage: pre-processing of the text of the articles was done (removing noise and stopwords). The second stage was the semantic operation, which was used to discover knowledge and extract meanings from Gibbs sampling. Then the data were modeled using LDA (Latent Dirichlet Allocaon). LDA assumes that each word in a document is generated in two steps. First, assuming that each document has its own topic distribution, a topic is randomly selected based on the topic distribution of the document. In the next step, assuming that each topic has its own word distribution, a word is randomly drawn from the word distribution of the topic selected in the previous step. By repeating these two steps word by word a document is created

Basically, LDA reduces the super dimensionality of the researcher's data from words to topics, based on the repetition of words in the same document, similar to cluster analysis or principal component analysis. According to the LDA algorithm, a set of documents is defined as Scholar-Context and words are defined as topics. Finally, by considering the marginal probability production of each researcher's article in the collection, the probability of the process of building a corpus is determined.

In step 3, after semantic extraction and topic extraction, in this step keywords are generated as an explosive topic based on ranking and high probability. These topics are recommended as data science research areas (**Figure 1**) [25].

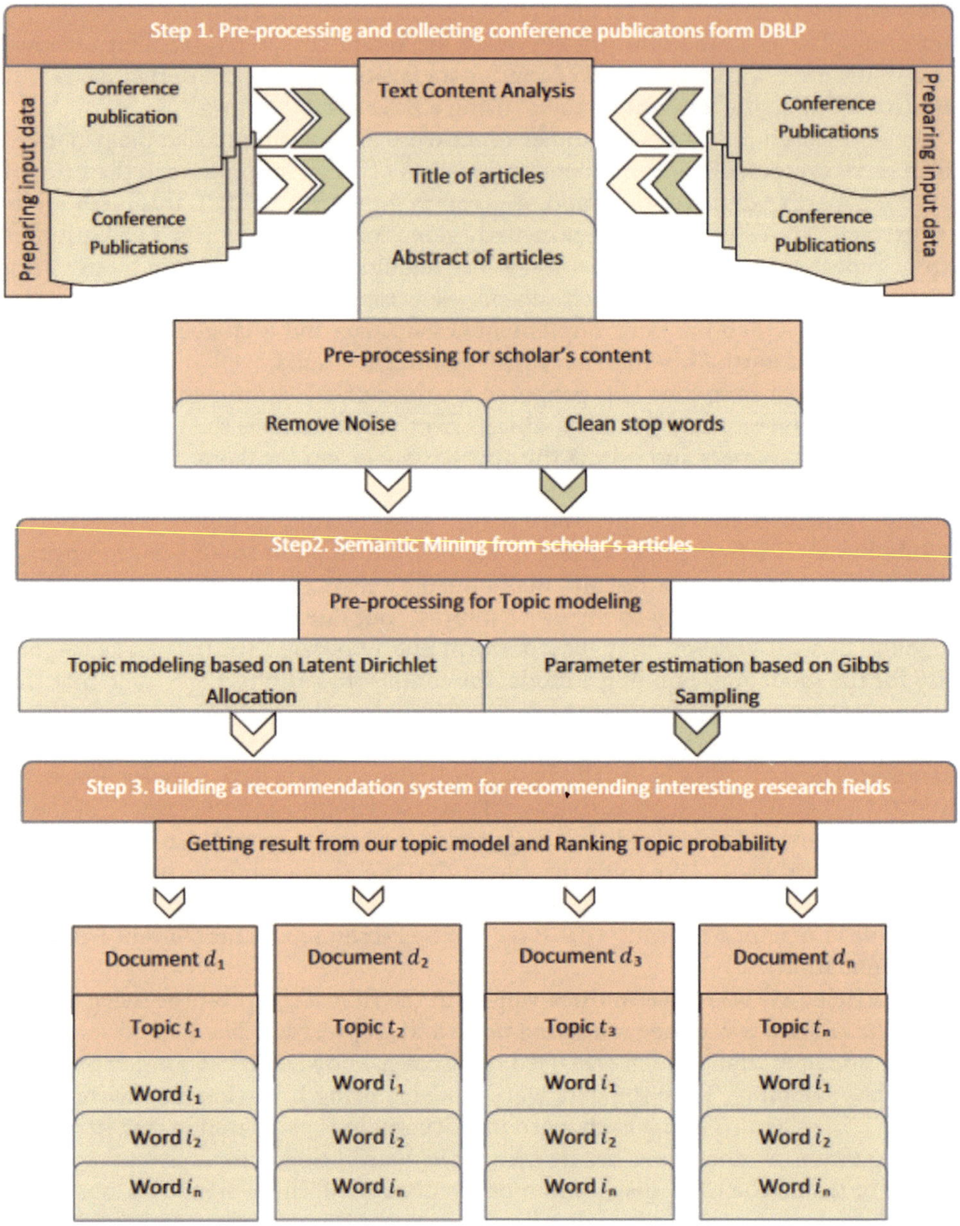

Figure 1.
The research framework for recommending research field and semantic mining from scholar's paper [25].

2.6 A recommender system in the Big Data platform

2.6.1 Hadoop platform

In reference [30] paper presented a method based on the background information of the papers in the Hadoop platform. They collected a dataset of conference papers from 2013 to 2015 of the IEEE Digital Library in the three domains of computer, electrical, and mechanical using a crawler that received the URL of each paper as crawler input and an XML file containing the attributes Fields such as the title of the paper, the abstract of the paper, the keywords of the paper were extracted as output, and the keywords part included the keywords that the author considered for the paper, the terms that IEEE considered, the indicators controlled in INSPEC And the non-controlled indicators are in INSPEC.

Their proposed system included two steps: (1) identifying the scope of papers and (2) recommending similar papers.

1. Identification of the scope of the papers

   Each paper profile contains sentences and words from the title, abstract and keywords that these words must be converted into a language that can be understood by the processor, therefore these words must be converted into code. In these sentences, there are letters such as is, are, when, etc., which are called inhibitory words and are also present in all papers, as a result, these letters must be removed, after removing these letters, the words must be rooted, then these words are weighted and Convert to code. Keywords and other terms specified in the profile and words that are mentioned more than twice in the abstract are considered as the domain vector of the paper. For 80% of the papers, their domains were trained, and for the remaining 20%, using the similarity function between the domain vector of the new papers and the domain vector of the dataset papers, it was detected and stored in the same domain.

   I. Remove StopWords and Finding Roots

   First, we create a class named RemoveStopWord, which includes three subclasses named deleteStopWord, Stem, and main, then we explain the function of each subclass.

   deleteStopWord:

   This subclass first receives a list of inhibitor words as a collection and converts this collection into an array list.

   Stem:

   This subclass is the implementation of Porter's algorithm, which first shows a word as input and the root of that word as output. This algorithm can be added to Java libraries by using library functions.

   Main:

   This subclass consists of two parts:

   A. Create a list of blocking words

      In this section, the list of blocking words from the input file is received as input and a set of blocking words is created.

   B. Converting a dataset containing words into an equivalent dataset where the words are the roots of the words.

      The system data set file, which was described in the method of collecting and creating it at the beginning of the chapter, is received as input. This function converts the data set file to a data set containing the root words used in it.

II. Weighting words based on tf/idf

This section is implemented by the TFID class, which uses mapping and reduction methods to implement the mapping-reducing model (explained in the previous chapter). Since word weighting is based on ngram, we need two helper functions which are defined as follows. In that research, the researchers considered $n = 2$, which is received as an input along with a sentence of words of each paper profile, the program divides the sentences into two words and shows them as output.

TFID production mapping

Each line of the new data set has a line number, which is considered as a record number in the mapping stage, and the content section (including title words, abstract and keywords) is considered as "value".

The output of this step is key-value. The key is the extracted bigrams (TFID column) and the value containing the calculated TFID number (which is currently considered to be 1) so that the original TFID value can be calculated by adding the numbers of one (1) corresponding to each bigram in the reduction stage.

Reduction in TFID production

In this stage, the input of the program is the output of the mapping stage, which produces TFID, bigrams of the textual part of each paper (title, abstract and keywords) and shows them as output.

If the weighted papers are part of 80% of the educational collection, the next step III is implemented, otherwise the papers are part of the educational collection, step IV is implemented.

III. Utility matrix

After weighting all the textual parts of the papers, these weights along with the bigrams and the ID of each paper enter in a useful matrix which is created in the form of key-value as intermediate Hadoop files so that the key or line number is similar to the ID of the paper. And the value or column is the same bigrams are stored, and the numbers inside the matrix are the tf/idf weight of that bigram in that paper and its range. This 80% is the training part of the data set.

IV. Test set

We considered 20% of the data set for the test set that the scope of the papers is recognized based on the similarity between the keywords of the papers and the scope of the papers that have been specified and included in the usefulness matrix. First, the profiles of the papers are returned to step I and II to remove the blocking words, root and weight them, and in step III, considering that these profiles are part of the test set, they go to step V to detect the domain.

V. Domain recognition

To detect the scope of the papers, first, we obtain the similarity of the vector obtained from the weighting of the words with the scope of the papers that were stored in the usefulness matrix (we consider the bigrams in the columns of the matrix as the scope of each domain) using cosine similarity. , which if there is a high similarity, the domain of the mentioned paper is determined and stored in that domain. To optimize the system, this step is merged with step 2.

All the previous steps were related to building the dataset and now we enter the part of our recommender system. If the incoming user is new and is using this system for the first time, he will go to the next step, otherwise, he will go to step 2-II.

The system randomly suggests some papers from each domain, and by choosing one of them, the user's working domain is determined for the system and the system saves it in the user's profile. Then it goes to the next step.

2. Building a recommender system and recommending papers

For this section, a class named recommender has been created, which uses four other classes named MapperJoin, MapperRecommender, ReducerJoin, and Reducer for mapping-reduction steps. In the following, the mode of operation, input and output are discussed.

This part is the mapping and reduction operation that takes place in two separate stages and is responsible for setting up the Hadoop driver.

I. Mapping MapperRecommender

This class, which performs the mapping operation, performs a simple mapping operation in order to perform the desired operation in a suitable form in the reduction phase, which is the input of the TFID table, which includes the bigrams of the title, abstract, and keywords.

II. Reducer mapping

In this mapping, by receiving the output of the mapping stage of the previous stage, it produces the output that provides the possibility of checking the similarity of the papers entered as input and the papers in the dataset. These papers can be recommended to the user who imported the original papers.

III. Mapping MapperJoin

This mapping performs the mapping operation to check the similarity between the papers. At this stage, the goal is to put together the same bigrams (with the same key) so that papers with the same bigram are put together, and this is necessary because of the suggestion of papers. Because the papers that have the same bigram are the same papers that have the same words in the title, abstract, keywords or the same or common key area.

IV. Reduce ReducerJoin

At this stage, similar papers based on the degree of similarity between the papers that the user has chosen for the first time or chosen as his work domain with the papers of the dataset are suggested using cosine similarity.

At this stage, the recommended paper along with its domain is saved in the user's profile for further recommendations, and finally the work of the recommended recommender system is finished.

Finally, the researchers [30] used the two criteria of user satisfaction and accuracy provided in reference [7] for evaluation, which is shown below in formula 3:

$$\text{SAT} = \frac{\text{Number of correct research papers}}{\text{Total number of recommended papers}}$$

$$\text{ACC} = \frac{\text{Recommended by SAT} + \text{Not recommended by UNSAT}}{\text{Total number of papers saved for a specific subject}} \tag{3}$$

According to the above formula, SAT shows the ratio of correct research papers which are consistent with the user domain, to the number of recommended papers and ACC shows the proportion of correct or incorrect research papers consistent with the user domain to the total number of saved papers for a specific domain.

2.6.2 Mahout platform

Here in reference [31], the implementation is done in four phases: the first phase is to extract ideas from the paper, the second phase is to extract the author's work area from the written paper, the third phase is to create a profile for the author (user), and finally, the fourth phase is to recommend the user's favorite paper. Which will be explained in the following four phases.

1. Phase one: extracting ideas from the paper

   In this part, two methods are used to get the idea of the paper. In the first part of the title

   The paper and in the second part of the paper abstract and title are considered to extract the idea of the paper.

   I. Extracting the idea of the paper from the abstract

   In the following code, the abstract part of the paper is used to extract ideas using the dataset containing the paper's specifications in XML form, and the cue words are removed using the RemovecueWords function, and then the idea of the paper is extracted using the getIdea function. The following shows how these functions work.

   RemovecueWords:

   A class named RemovecueWords is created. Which first receives a list of reference words as a set and considers the sentences of the abstract that contain reference words, removes these words from the sentences and sends the sentence without reference words to the output.

getIdea:

In the getIdea function class, it first tags (POS tagger) the words extracted from the previous section, and searches for ideas with consecutive words with the name and attribute tag. For tagging, MaxenTagger is used, which is added to the Java library as a jar file, and as input file, which is a stemmer configuration file, is provided to MaxenTagger.

II. Extracting the idea of the paper from the body of the paper

At this stage, considering that the purpose of extracting ideas is to calculate the similarity of the title with the body of the paper, we need to extract the title and body of the paper and perform a series of pre-processing to extract ideas. In this part, the text of the paper is used (the same text as the existing references). Generally, sentences are used that are similar to the title because the title contains important and key terms and sentences that have these terms and are considered similar because they contain important terms here via TF The number of repetitions of words with the highest frequency is found, that is, the sentences that have the most similarity to the title are extracted. The title part of the paper is extracted from the input files for future processing.

Remove common words:

In this section, a list of words that are usually repeated the most in the text of the paper is formed, and then the sentences that contain these words are deleted by this function. Then, the similarity of each sentence from the abstract in the previous step is considered with the title, and each sentence that has the highest frequency in terms of similarity, and if it reaches the set threshold limit, that sentence is recorded as an paper idea in the user's profile.

2. The second phase: extracting the topic and scope of the paper

At this stage, two methods are used to extract the topic of the paper. The first method is to use the title of the paper and the references section, and the second method is to use the abstract of the paper.

I. Extracting the scope of work from the title and references

In this part, we need to find the root, remove the inhibitory words and separate the sentences in the text according to the punctuation marks, and those sentences that have a high frequency between the punctuation marks after pre-processing are stored as topic in HDFS and user profile.

A. Title

Finding blocking words and finding roots

First, we create a class called Remove stop word, which includes three subclasses: LinkedHashSet, deletestopword, and stem subclasses, which are as follows:

a. LinkedHashSet:

This subclass is used to generate stopwords

b. deleteStopWord:

This subclass takes the inhibitory words in the previous section as input and puts them in the array, considers the inhibitory words, breaks the sentence from the inhibitory words section, and finally removes the inhibitory words.

c. Stem:

This subclass is the implementation of Porter's algorithm, which takes the broken sentences in the previous section and receives their words as input and shows their roots as output. Rooting through the following code using the porter root finder jar file performs the rooting operation

B. Referrals

In addition to the work of the previous part, punctuation marks are used in the references. We receive the punctuation marks in the array and break the references from the punctuation marks part and find the root and finally remove these marks. At the end, after finding the roots of the words of the broken sentences, we find the frequency of these words in the abstract and consider the words with the highest frequency as the author's work area.

II. Extracting the work area from the abstract and text of the paper

It finds the words that have the highest frequency in the body and the abstract, or the words that have the most repetition together, and at the end, it deletes similar words and combines other words.

3. The third phase: creating a user profile

According to the information obtained from the previous sections and the information in the data set such as the title, abstract, keywords, scope and idea of the paper, the feature vector is formed to create a profile.

4. The fourth phase: building a recommender system and recommending papers

At first, according to the feature vectors built in the previous part, Kisenosi similarity is used to detect the similarity of each feature in the author's profile with the candidate paper, and to build the recommendation system, the Mahout platform is used due to its simplicity in its implementation and also Most of the library functions are ready, the code below the implementation steps of the recommendation part. For example, the author's field of work is extracted by looking at the title of the paper and finding the inhibitory word in it, and breaking the title of the paper into two sentences from the inhibitory word, then finding the roots of the words of each sentence, and finally the frequency of each It finds an abstract of the words and registers it as the author's field of work in the author's profile and presents the most similar paper to the user according to the feature vector in the user's profile and using cosine similarity. Finally, the researchers used recall criteria and DCG for evaluation, which compared to other researches, the proposed method has provided an acceptable performance.

3. Conclusion

With the increasing number of data in DLS, they, like other businesses, have taken advantage of recommender systems and try to personalize recommendations to users by using the textual information of papers. This chapter examines text-aware recommender systems, recommending papers for referring data to draft papers, personalized recommendations, extracting user interests and behavior and analyzing them for personalized recommendations and extracting the scope of papers, as well as paper recommender systems using It has provided big data platforms. The recommender systems proposed in this chapter used the textual information of papers to create paper profiles, according to which it is easy to recognize the scope of new papers and the scope of work of each user, and provide satisfactory recommendations according to the scope and interests of the user. Also, some researches also formed profiles for users according to the papers and recommended papers using the CF method.

Author details

Nasrin Jokar[1*], Mehra Esfandiari[1], Shima Aghamirzadeh[2] and Hossein Hatami[3]

1 B2CTech Research Group, Shiraz, Iran

2 Department of Electrical and Computer Engineering, Payam Noor, Ramhormoz, Iran

3 Department of Electrical and Computer Engineering, Safashahr Branch, Islamic Azad University, Safashahr, Iran

*Address all correspondence to: nasrin.jokar@gmail.com

References

[1] Yu M-C, Jim Wu Y-C, Alhalabi W, Kao H-Y, Wu W-H. ResearchGate: an effective altmetric indicator for active researchers? Computers in Human Behavior. 2016;**55**:1001-1006. DOI: 10.1016/j.chb.2015.11.007

[2] Beel J, Langer S, Genzmehr M, Gipp B, Breitinger C, Nürnberger A. Research paper recommender system evaluation: a quantitative literature survey. In: Proceedings of the International Workshop on Reproducibility and Replication in Recommender Systems Evaluation, 2013. pp. 15-22. DOI: 10.1145/2532508.2532512

[3] Beel J, Gipp B, Langer S, Breitinger C. Paper recommender systems: a literature survey. International Journal on Digital Libraries. 2016;**17**(4):305-338. DOI: 10.1007/s00799-015-0156-0

[4] Caragea C, Silvescu A, Mitra P, Lee Giles C. Can't see the forest for the trees? a citation recommendation system. In: Proceedings of the 13th ACM/IEEE-CS joint conference on Digital libraries, 2013. pp. 111-114. DOI: 10.1145/2467696.2467743

[5] Rokach L, Mitra P, Kataria S, Huang W, Giles L. A supervised learning method for context-aware citation recommendation in a large corpus. INVITED SPEAKER: Analyzing the Performance of Top-K Retrieval Algorithms 2013: 2013

[6] Hong K, Jeon H, Jeon C. UserProfile-based personalized research paper recommendation system. In: 2012 8th International Conference on Computing and Networking Technology (INC, ICCIS and ICMIC), IEEE, 2012. pp. 134-138

[7] Hong K, Jeon H, Jeon C. Personalized research paper recommendation system using keyword extraction based on userprofile. Journal of Convergence Information Technology. 2013;**8**(16):106

[8] Wang Y, Liu J, Dong XL, Liu T, Huang YL. Personalized paper recommendation based on user historical behavior. In: CCF International Conference on Natural Language Processing and Chinese Computing. Berlin, Heidelberg: Springer; 2012. DOI: 10.1007/978-3-642-34456-5_1

[9] Champiri ZD, Shahamiri SR, Salim SSB. A systematic review of scholar context-aware recommender systems. Expert Systems with Applications. 2015;**42**(3):1743-1758. DOI: 10.1016/j.eswa.2014.09.017

[10] He Q, Pei J, Kifer D, Mitra P, Giles L. Context-aware citation recommendation. In: Proceedings of the 19th international conference on World wide web, 2010. pp. 421-430. DOI: 10.1145/1772690.1772734

[11] Huang W, Wu Z, Liang C, Mitra P, Lee Giles C. A neural probabilistic model for context based citation recommendation. In: Twenty-ninth AAAI conference on artificial intelligence. 2015. DOI: 10.1609/aaai.v29i1.9528

[12] Rokach L, Mitra P, Kataria S, Huang W, Giles L. A supervised learning method for context-aware citation recommendation in a large corpus. INVITED SPEAKER: Analyzing the Performance of Top-K Retrieval Algorithms 1978:1978

[13] Sun Y, Ni W, Men R. A personalized paper recommendation approach based on web paper mining and reviewer's interest modeling. In: 2009 International Conference on Research Challenges in

Computer Science. IEEE, 2009. pp. 49-52. DOI: 10.1109/ICRCCS.2009.76

[14] Lu Y, He J, Shan D, Yan H. Recommending citations with translation model. In: Proceedings of the 20th ACM International Conference on Information and Knowledge Management. Oct 2011. pp. 2017-2020. DOI: 10.1145/2063576.2063879

[15] Lee J, Lee K, Kim JG. Personalized academic research paper recommendation system. arXiv preprint arXiv:1304.5457 2013. DOI: 10.48550/arXiv.1304.5457

[16] Liu F, Yu C, Meng W. Personalized web search by mapping user queries to categories. In: Proceedings of the eleventh international conference on Information and knowledge management. 2002. pp. 558-565. DOI: 10.1145/584792.584884

[17] Huang W, Kataria S, Caragea C, Mitra P, Lee Giles C, Rokach L. Recommending citations: translating papers into references. In: Proceedings of the 21st ACM international conference on Information and knowledge management. 2012. pp. 1910-1914. DOI: 10.1145/2396761.2398542

[18] McNee SM, Albert I, Cosley D, Gopalkrishnan P, Lam SK, Rashid AM, et al. On the recommending of citations for research papers. In: Proceedings of the 2002 ACM Conference on Computer Supported Cooperative Work. Nov 2002. pp. 116-125. DOI: 10.1145/587078.587096

[19] Medić Z, Šnajder J. An empirical study of the design choices for local citation recommendation systems. Expert Systems with Applications. 2022; **200**:116852. DOI: 10.1016/j.eswa.2022.116852

[20] Ali Z, Kefalas P, Muhammad K, Ali B, Imran M. Deep learning in citation recommendation models survey. Expert Systems with Applications. 2020;**162**: 113790. DOI: 10.1016/j.eswa.2020.113790

[21] Wang G, Wang HR, Yang Y, Xu DL, Yang JB, Yue F. Group article recommendation based on ER rule in Scientific Social Networks. Applied Soft Computing. 2021;**110**:107631. DOI: 10.1016/j.asoc.2021.107631

[22] Chaudhuri A, Sarma M, Samanta D. SHARE: Designing Multiple Criteria-Based Personalized Research Paper Recommendation System. Information Sciences. 2022. DOI:10.1016/j.ins.2022.09.064

[23] Chaudhuri A, Samanta D, Sarma M. Modeling user behaviour in research paper recommendation system. arXiv preprint arXiv:2107.07831. 2021. DOI: 10.48550/arXiv.2107.07831

[24] Patel K, Caragea C, Wu J, Giles CL. Keyphrase extraction in scholarly digital library search engines. In: International Conference on Web Services. Cham: Springer; Sep 2020. pp. 179-196. DOI: 10.1007/978-3-030-59618-7_12

[25] Jelodar H, Wang Y, Xiao G, Rabbani M, Zhao R, Ayobi S, et al. Recommendation system based on semantic scholar mining and topic modeling on conference publications. Soft Computing. 2021;**25**(5):3675-3696. DOI: 10.1007/s00500-020-05397-3

[26] Strohman T, Bruce Croft W, Jensen D. Recommending citations for academic papers. In: Proceedings of the 30th annual international ACM SIGIR conference on Research and development in information retrieval. 2007. pp. 705-706. DOI: 10.1145/1277741.1277868

[27] Lakhanpal S, Gupta A, Agrawal R. Towards Extracting Domains from Research Publications. In: MAICS. 2015. pp. 117-120

[28] Lakhanpal S, Gupta A, Agrawal R. Discover trending domains using fusion of supervised machine learning with natural language processing. In: 2015 18th International Conference on Information Fusion (Fusion), IEEE. 2015. pp. 893-900

[29] Lakhanpal S, Gupta A, Agrawal R. On discovering most frequent research trends in a scientific discipline using a text mining technique. In: Proceedings of the 2014 ACM Southeast Regional Conference. 2014. pp. 1-4. DOI: 10.1145/2638404.2638528

[30] Jokar N, Honarvar AR, Esfandiari K. A contextual information based scholary paper recommender system using big data platform. Journal of Fundamental and Applied Sciences. 2016;**8**(2): 914-924. DOI: 10.4314/jfas.v8i2s.144

[31] Aghamirzad S, Honarvar AR, Jokar N. A paper recommender system based on user's profile in big data scholarly. Journal of Fundamental and Applied Sciences. 2016;**8**(2):941-955. DOI: 10.4314/jfas.v8i2s.150

Chapter 11

A Comparative Analysis of Search Algorithms for Solving the Vehicle Routing Problem

Oladimeji Samuel Sowole

Abstract

The Vehicle Routing Problem (VRP) is an extensively studied optimization challenge in operations research, applicable to logistics, transportation, and supply chain management. Its goal is to find optimal routes for vehicles, minimizing distance and maximizing customer satisfaction. Genetic algorithms, simulated annealing, and ant colony optimization are search algorithms commonly used to solve the VRP. This chapter provides a comparative analysis of these algorithms, highlighting their strengths and weaknesses. It introduces the VRP and its variants, along with associated challenges and constraints, and offers an overview of different search algorithms used for solving the problem, explaining their principles, advantages, and limitations. Real-world case studies showcase successful applications of these algorithms in package delivery, waste collection, and emergency response. Additionally, the chapter explores key factors influencing algorithm performance, including problem size, complexity, and parameters. It concludes by providing recommendations for selecting appropriate algorithms for different VRP instances. By providing a comprehensive understanding of search algorithms for the VRP, this chapter enables readers to make informed decisions when addressing similar optimization problems in practical scenarios.

Keywords: vehicle routing problem (VRP), search algorithms, optimization, metaheuristics, ant colony optimization (ACO), particle swarm optimization (PSO), genetic algorithms (GA), tabu search, simulated annealing, heuristics, routing strategies

1. Introduction

The Vehicle Routing Problem (VRP) is a well-known optimization problem in operations research, with significant applications in logistics, transportation, and supply chain management. It involves determining the optimal routes for a fleet of vehicles to serve a set of customers while minimizing costs and maximizing customer satisfaction. Efficiently solving the VRP has substantial implications for reducing transportation costs, improving resource utilization, and enhancing overall operational efficiency [1–3].

1.1 Problem statement and objectives

The VRP can be formally defined as follows:
Given:

1. A set of customers, denoted by $C = \{c_1, c_2, \ldots, c_n\}$

2. A depot, denoted by D

3. A fleet of vehicles, denoted by $V = \{v_1, v_2, \ldots, v_m\}$

4. Each customer c_i has a demand d_i, representing the quantity of goods to be delivered

5. Each vehicle v_i has a capacity q, representing the maximum amount it can carry

6. The distance between any two locations i and j is represented by $d(i,j)$.

The objective of the VRP is to find:

- For each vehicle v_i, a route denoted by R_i, starting and ending at the depot D

- The sequence of customers to be visited along each route

The objectives of the VRP are to:

- Minimize the total distance traveled by all vehicles

- Minimize the number of vehicles used

- Balance the workload among the vehicles

- Satisfy the demand of each customer.

1.2 Definition and variants of VRP

The VRP encompasses various problem variants, each introducing additional constraints or objectives. Some common variants of the VRP include:

1.2.1 Capacitated VRP (CVRP)

In Capacitated VRP (CVRP), each vehicle has a limited capacity, and the total demand served by each vehicle must not exceed its capacity [3]. **Figure 1** gives an illustration of CVRP. Mathematically, CVRP can be formulated as follows:
Minimize:

$$\sum d(i,j) * x(i,j) \tag{1}$$

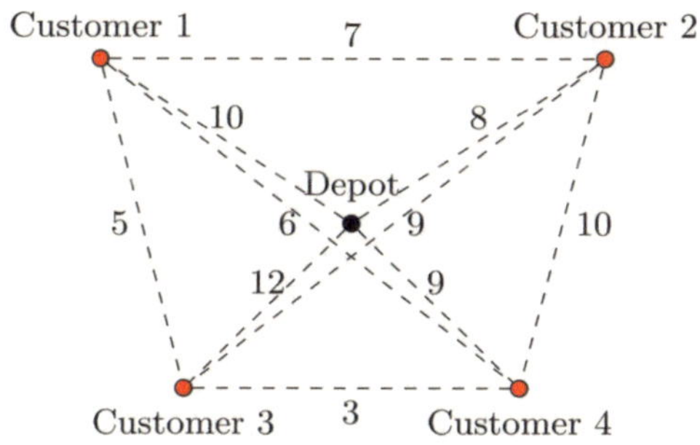

Figure 1.
Capacitated vehicle routing problem (CVRP) illustration.

Subject to:

1. $\sum x(i,j) = 1$, for all $i \in C, j \in C, i \neq j$

2. $\sum x(j,i) = 1$, for all $i \in C, j \in C, i \neq j$

3. $\sum x(i,j) \leq m * y(i)$, for all $i \in C, j \in C, i \neq j$

4. $\sum x(j,i) \leq m * y(i)$, for all $i \in C, j \in C, i \neq j$

5. $\sum y(i) = m$

6. $\sum d(i) * y(i) \leq q$, for all $i \in C$

7. $x(i,j) \in \{0,1\}$, for all $i \in C, j \in C, i \neq j$

8. $y(i) \in \{0,1\}$, for all $i \in C$.

Where:

- $x(i,j)$ is a binary decision variable denoting whether customer i is assigned to customer j in the solution
- $y(i)$ is a binary decision variable denoting whether customer i is visited in the solution
- $d(i,j)$ represents the distance between customers i and j
- q is the capacity of each vehicle
- m is the total number of vehicles.

1.2.2 Vehicle routing problems with time windows (VRPTW)

For Vehicle Routing Problems with Time Windows (VRPTW), customers have specified time windows during which they can be served, and the vehicles must respect these time constraints [4]. **Figure 2** gives a schematic illustration of VRPTW.

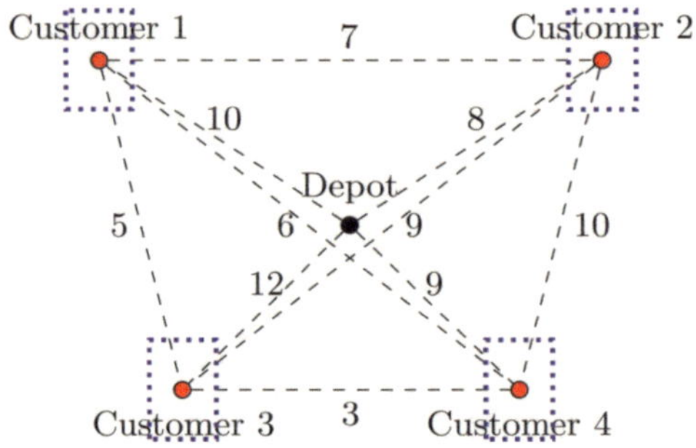

Figure 2.
Vehicle routing problem with time windows (VRPTW) illustration.

Mathematically, Vehicle Routing Problem with Time Windows (VRPTW) can be formulated as follows:

Minimize:

$$\sum d(i,j) * x(i,j) \tag{2}$$

Subject to:

1. $\sum x(i,j) = 1$, for all $i \in C, j \in C, i \neq j$

2. $\sum x(j,i) = 1$, for all $i \in C, j \in C, i \neq j$

3. $\sum x(i,j) \leq m * y(i)$, for all $i \in C, j \in C, i \neq j$

4. $\sum x(j,i) \leq m * y(i)$, for all $i \in C, j \in C, i \neq j$

5. $\sum y(i) = m$

6. $\sum d(i) * y(i) \leq q$, for all $i \in C$

7. $a(i) \leq t(i) \leq b(i) - d(i,j) - M * (1 - x(i,j))$, for all $i \in C, j \in C, i \neq j$

8. $x(i,j) \in \{0,1\}$, for all $i \in C, j \in C, i \neq j$

9. $y(i) \in \{0,1\}$, for all $i \in C$.

Where:

- $x(i,j)$ is a binary decision variable denoting whether customer i is assigned to customer j in the solution
- $y(i)$ is a binary decision variable denoting whether customer i is visited in the solution
- $d(i,j)$ represents the distance between customers i and j
- q is the capacity of each vehicle

- m is the total number of vehicles
- $a(i)$ and $b(i)$ represent the start and end of the time window for customer i, respectively
- $t(i)$ represents the arrival time at customer i
- M is a large positive constant.

1.3 Main challenges and constraints

The VRP presents several challenges and constraints that need to be considered in the solution process. Some of the main challenges and constraints include:

1. Capacity Constraint: Each vehicle has a limited capacity, and the total demand served by each vehicle must not exceed its capacity. Mathematically, this constraint can be represented as: $\sum d(i,j) * y(i,j) \leq q$, for all $i \in C, j \in C, i \neq j$.

2. Time Window Constraint: In VRPTW, customers have specified time windows during which they can be served. The vehicles must respect these time constraints and ensure that the arrival time at each customer falls within its time window. This constraint can be expressed as: $a(i) \leq t(i) \leq b(i) - d(i,j) - M * (1 - x(i,j))$, for all $i \in C, j \in C, i \neq j$. Where $a(i)$ and $b(i)$ represent the start and end of the time window for customer i, respectively, $t(i)$ is the arrival time at customer i, $d(i,j)$ is the travel time between customers i and j, and M is a large positive constant.

3. Routing Constraints: Each vehicle must start and end its route at the depot, and it should visit each customer exactly once. Additionally, the routes should not contain subcycles or disconnected components.

4. Objective Function: The objective is to minimize the total distance traveled by all vehicles while satisfying the demand of each customer and respecting the capacity and time window constraints.

Solving the VRP considering these challenges and constraints requires the application of efficient search algorithms. In the following section, we will explore and analyze different search algorithms, namely genetic algorithms, simulated annealing, and ant colony optimization, which have been widely used to tackle the VRP.

2. Search algorithms for VRP

2.1 Genetic algorithms

Genetic algorithms (GAs) are population-based search algorithms inspired by the process of natural evolution. They utilize genetic operators, such as selection,

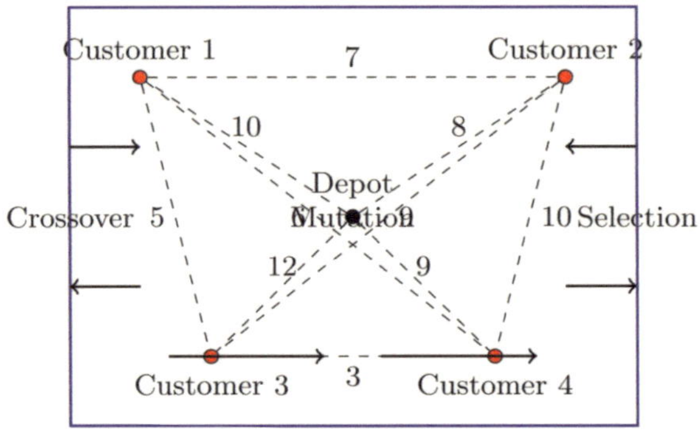

Figure 3.
Genetic algorithm (GA) for vehicle routing problem (VRP) illustration.

crossover, and mutation, to iteratively evolve a population of candidate solutions toward better solutions. GAs have been widely applied to solve various optimization problems, including the VRP [5, 6]. We give an illustration of GA in **Figure 3**.

2.1.1 Genetic representation and operators

In the context of the VRP, a common representation of a solution is a chromosome, which encodes the sequence of customers to be visited by a vehicle. The genetic operators used in GAs for the VRP are as follows:

1. Initialization: Generate an initial population of random feasible solutions.

2. Fitness evaluation: Evaluate the fitness of each solution based on the objective function (e.g., total distance traveled).

3. Selection: Select a subset of solutions from the population based on their fitness values, giving preference to better solutions.

4. Crossover: Combine pairs of selected solutions to create offspring solutions. This involves exchanging segments of routes between parents.

5. Mutation: Randomly modify some elements (e.g., customer sequence) in the offspring solutions to introduce diversity.

6. Replacement: Replace some solutions in the population with the newly created offspring solutions.

7. Termination: Repeat the above steps until a termination condition is met (e.g., a maximum number of generations or a convergence criterion).

The genetic operators allow exploration of the solution space by preserving good characteristics from the parent solutions and introducing variation through crossover and mutation.

2.2 Fitness evaluation

The fitness function in GAs for the VRP measures the quality of a solution based on the objectives of minimizing the total distance traveled, balancing the workload among vehicles, and satisfying the capacity constraints. The fitness evaluation typically involves calculating the total distance traveled by each vehicle, considering the sequence of customers visited.

2.2.1 Mathematical formulation

Given a solution represented by a chromosome, the fitness function can be defined as:

$$Fitness = (\alpha * TotalDistance) + (\beta * WorkloadImbalance) + (\gamma * CapacityViolation) \quad (3)$$

where:

$\alpha, \beta,$ and γ are weight coefficients that balance the importance of each objective. The total distance can be calculated by summing the distances between consecutive customers visited by each vehicle. Workload imbalance can be quantified by measures such as the standard deviation of the vehicle workloads or the maximum workload difference. Capacity violation accounts for the amount by which the demand exceeds the vehicle capacity.

2.2.2 Advantages and limitations

The advantages of using genetic algorithms for the VRP include the:

- Ability to explore a large solution space efficiently.
- Ability to handle different problem variants and constraints.
- Potential to find high-quality solutions.

However, GAs also have some limitations:

- Computational complexity increases with problem size, limiting their applicability to large-scale instances.
- Difficulty in striking a balance between exploration and exploitation.
- Selection of appropriate parameter values can be challenging.

2.3 Simulated annealing

Simulated annealing (SA) is a stochastic search algorithm that emulates the annealing process in metallurgy. It uses a probabilistic acceptance criterion to accept worse solutions early in the search process and gradually focuses on improving the objective function [7, 8]. **Figure 4** gives a schematic illustration of SA.

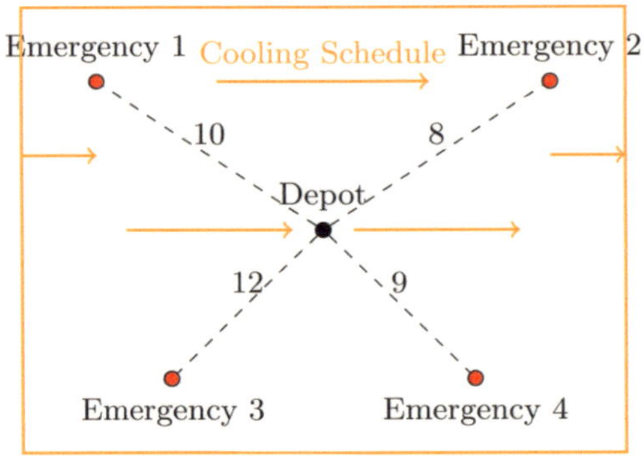

Figure 4.
Simulated annealing (SA) for vehicle routing problem (VRP) illustration.

2.3.1 Annealing schedule and neighborhood structure

In SA for the VRP, an initial solution is generated, and a temperature parameter is set. The algorithm iteratively explores the solution space by perturbing the current solution and accepting new solutions based on a probability function that depends on the objective function difference and the current temperature [7, 8]. The neighborhood structure defines the set of feasible moves from the current solution. In the VRP, common neighborhood structures include swapping customers between routes or inserting a customer from one route into another route while maintaining the capacity and time window constraints.

2.3.2 Acceptance criteria

The acceptance criterion in SA determines whether to accept a new solution based on its objective function value and the current temperature. Initially, the algorithm allows a higher probability of accepting worse solutions, which facilitates exploration. As the temperature decreases, the algorithm gradually becomes more selective and focuses on improving the objective function.

2.3.3 Mathematical formulation

Given the current solution with objective function value F(current) and a new solution with objective function value F(new), the acceptance probability can be defined as:

$$P = \frac{\exp((F(current) - F(new))}{T}, \tag{4}$$

where T is the current temperature.

The acceptance probability allows the algorithm to accept worse solutions with a higher probability at the beginning, gradually reducing the probability as the temperature decreases.

2.3.4 Advantages and limitations

Advantages of simulated annealing for the VRP include:

- Ability to escape local optima by allowing moves that worsen the objective function.

- Flexibility in exploring the solution space through the acceptance probability.

- Effective for problems with a complex search landscape.

However, SA also has some limitations:

- Dependence on parameter tunings, such as the initial temperature and the cooling schedule.

- Convergence to suboptimal solutions if the cooling schedule is not appropriately chosen.

- Computationally intensive for large problem instances.

2.4 Ant colony optimization

Ant Colony Optimization (ACO) is a metaheuristic algorithm inspired by the foraging behavior of ants. It utilizes pheromone-based communication and stigmergy to iteratively construct high-quality solutions to the VRP. **Figure 5** is a schematic illustration of ACO.

2.4.1 Pheromone-based communication and stigmergy

In ACO for the VRP, artificial ants construct solutions by iteratively moving from one customer to another, depositing pheromones along their paths [9]. The concentration of pheromone on an edge represents the desirability of that edge. Ants probabilistically choose the next customer to visit based on the pheromone levels and heuristic information (e.g., distance) of the neighboring customers. Pheromone evaporation and reinforcement mechanisms ensure the dynamic update of pheromone trails. Evaporation reduces the pheromone levels over time to avoid convergence to suboptimal solutions, while reinforcement strengthens the pheromone trail of good-quality solutions found by the ants [9, 10].

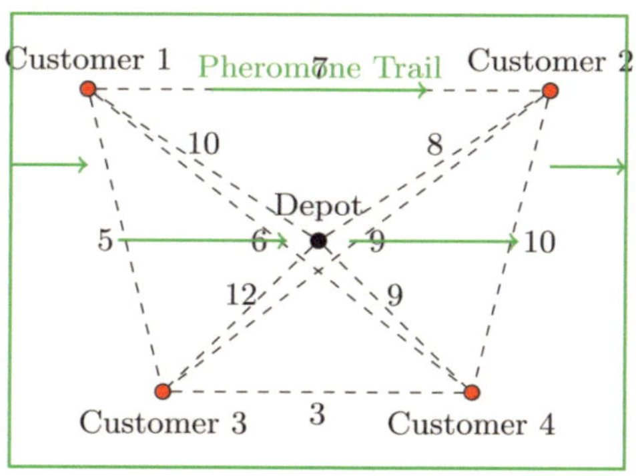

Figure 5.
Ant colony optimization (ACO) for vehicle routing problem (VRP) illustration.

2.4.2 Construction of routes and pheromone update

In ACO, each ant constructs a complete set of routes by iteratively selecting customers to visit based on the pheromone and heuristic information. The pheromone update is performed by evaporating the existing pheromone and depositing the pheromone on the edges traversed by the ants based on the quality of the solution found.

2.4.3 Mathematical formulation

The pheromone update rule can be expressed as:

$$\tau(i,j) = (1-\rho) * \tau(i,j) + \sum(\Delta\tau k(i,j)) \tag{5}$$

Where:
$\tau(i,j)$ represents the pheromone level on the edge (i,j), ρ is the evaporation rate, and $\Delta\tau k(i,j)$ represents the pheromone deposit made by ant k on the edge (i,j).
The pheromone deposit $\Delta\tau k(i,j)$ can be defined as:

$$\Delta\tau k(i,j) = \frac{Q}{Lk}, \tag{6}$$

where Q is a constant representing the pheromone quantity deposited by each ant and Lk is the objective function value (e.g., total distance) of the solution constructed by ant k.

2.4.4 Advantages and limitations

Advantages of Ant Colony Optimization for the VRP include:

- Robustness and adaptability to dynamic problem instances and changes in the solution space.
- Effective exploration of the solution space through the probabilistic decision-making process.
- Ability to capture and exploit good solutions found by the ants through pheromone reinforcement.
- Scalability to handle large-scale instances of the VRP.

However, ACO also has some limitations:

- Convergence to suboptimal solutions if the algorithm gets trapped in local optima.
- Sensitivity to parameter settings, such as the evaporation rate and pheromone deposit constant.
- Difficulty in handling complex constraints, especially with multiple objectives.

2.5 Other search algorithms for VRP

In addition to Genetic Algorithms, Simulated Annealing, and Ant Colony Optimization, several other search algorithms have been applied to solve the VRP. Here are some notable examples:

2.6 Tabu search

Tabu Search is a metaheuristic algorithm that employs a memory-based search strategy to explore the solution space [11]. It maintains a short-term memory called the tabu list, which prohibits recently visited solutions from being revisited. Tabu Search iteratively explores the neighborhood of the current solution, allowing moves that improve the objective function or violate certain tabu conditions [11].

2.6.1 Particle swarm optimization

Particle Swarm Optimization (PSO) is a population-based optimization algorithm inspired by the social behavior of bird flocking or fish schooling. In PSO, candidate solutions (particles) move through the solution space based on their own best-known position and the best-known position of the swarm. PSO aims to converge to optimal solutions by iteratively updating the velocities and positions of the particles.

2.6.2 Iterated local search

Iterated Local Search (ILS) is a metaheuristic algorithm that combines local search with perturbation and diversification [12]. It starts with an initial solution and applies a local search algorithm to improve it. After reaching a local optimum, ILS introduces perturbations to escape the local optima and continues the search. The process of local search and perturbation is repeated iteratively to improve the solution quality [12].

2.6.3 Hybrid approaches

Hybrid approaches combine multiple search algorithms or integrate heuristic rules with optimization algorithms to enhance the performance of VRP solvers [13]. For example, a hybrid algorithm may incorporate a construction heuristic to generate initial solutions, followed by a local search algorithm or a metaheuristic to refine the solutions.

These are just a few examples of the numerous search algorithms that have been explored for solving the VRP. Each algorithm has its strengths and limitations, making them suitable for different problem instances and objectives. The choice of the search algorithm depends on the specific requirements and constraints of the VRP instance at hand.

In the following section, we will present case studies and examples of how these search algorithms have been applied to solve the VRP in real-world scenarios, highlighting their performance and applicability.

3. Case studies and examples

In this section, we will explore various case studies and examples of using search algorithms to solve the Vehicle Routing Problem (VRP) in different real-world scenarios. We will discuss the problem formulation and algorithm implementation, and evaluate the performance of the algorithms in each case.

3.1 Vehicle routing for package delivery

Package delivery is a common application of the VRP, where a fleet of vehicles is responsible for delivering packages to customers efficiently. Let us consider a case where there are multiple vehicles and a set of customers with specific demands and time windows.

3.1.1 Problem formulation

Consider a scenario where a package delivery company needs to optimize the routes for a fleet of vehicles to deliver packages to a set of customers. Each customer has a specific location, package demand, and delivery time window. The objective is to minimize the total distance traveled by vehicles while ensuring that all packages are delivered within their respective time windows and that the vehicle capacity is not exceeded. Let:

- Set of customers $C = \{c_1, c_2, \dots, c_n\}$ with demands $d_1, d_2, \dots, d_n$
- Set of vehicles $V = \{v_1, v_2, \dots, v_m\}$ with capacities $q_1, q_2, \dots, q_m$
- Distance matrix D representing the distances between customers and depots
- Location of each customer $c \in C$ represented by coordinates (x_c, y_c)
- Package demand for each customer $c \in C$ denoted by d_c
- Time window for each customer $c \in C$ specified as $[earliest_{time_c}, latest_{time_c}]$
- Vehicle capacity for each vehicle $v \in V$ denoted by q_v

Objective:

Minimize the total distance traveled by all vehicles while satisfying the package delivery time windows and vehicle capacity constraints.

The mathematical formulation for the Vehicle Routing Problem for Package Delivery can be represented as given below.

Now, let us denote the decision variables:

- $x_{cvc} = 1$ if vehicle v visits customer c, 0 otherwise.
- d_{cd} = Distance between customer c and customer d.
- t_{cv} = Time at which vehicle v arrives at customer c.

The mathematical formulation for the Vehicle Routing for Package Delivery problem can then be represented as:

Minimize:

$$\sum_{v \in V} \sum_{c \in C} \sum_{d \in C} (d_{cd} * x_{cvc} * x_{dv}) \tag{7}$$

Subject to:

1. $\sum_{v \in V} \sum_{c \in C} (x_{cvc} * d_c) \leq q_v$, for all $v \in V$ (Vehicle capacity constraint)

2. $\sum_{v \in V} (x_{cvc}) = 1$, for all $c \in C$ (Each customer is visited exactly once)

3. $\sum_{c \in C} (x_{cvc}) = 1$, for all $v \in V$ (Each vehicle visits exactly one customer)

4. $t_{cv} \geq \frac{t_c + d_c}{V_{min}}$, for all $v \in V, c \in C$ (Time window constraint—Arrival time at each customer must be after its earliest time window)

5. $t_{cv} \geq \frac{t_c + d_c}{V_{max}}$, for all $v \in V, c \in C$ (Time window constraint—Arrival time at each customer must be before its latest time window)

6. $\sum_{c \in C} (t_{cv} * x_{cvc}) \leq T_v$, for all $v \in V$ (Total time traveled by each vehicle must be within its maximum time limit)

7. $x_{cvc} \in \{0, 1\}$, for all $c \in C, v \in V$ (Binary variable indicating whether a vehicle visits a customer)

The objective function aims to minimize the total distance traveled by all vehicles. It sums up the distances between customers c and d, multiplied by the binary variables x_{cvc} and x_{dv}, which indicate whether a vehicle visits those customers.

The first constraint ensures that the demand of each customer does not exceed the capacity of the vehicle assigned to it. It sums up the demands d_c at each customer c visited by vehicle v, multiplied by the binary variable x_{cvc}, and enforces that it must be less than or equal to the capacity q_v of vehicle v.

The second constraint ensures that each customer is visited exactly once. It states that the sum of binary variables x_{cvc} over all vehicles v must be equal to 1 for each customer c.

The third constraint ensures that each vehicle visits exactly one customer. It states that the sum of binary variables x_{cvc} over all customers c must be equal to 1 for each vehicle v.

The fourth and fifth constraints enforce the time window constraints. They state that the arrival time $t_c v$ at each customer c by vehicle v must be after its earliest time window (t_c) and before its latest time window (t_c). The travel time d_c divided by the maximum and minimum vehicle speeds (V_{max} and V_{min}) is added to the customer's time window.

The sixth constraint ensures that the total time traveled by each vehicle, considering the arrival times at each customer is within its maximum time limit T_v.

Finally, the binary variables x_{cvc} are constrained to be either 0 or 1, indicating whether a vehicle visits a customer.

By solving this mathematical formulation using appropriate search algorithms, an optimized set of routes for the package delivery vehicles can be obtained, minimizing the total distance traveled while satisfying the capacity and time window constraints. Here, we show how to implement this with Genetic Algorithm.

3.1.2 Algorithm implementation: Genetic algorithm

Here's a step-by-step implementation of the Genetic Algorithm for solving the VRP for package delivery:

Step 1: Initialization

- Generate an initial population of solutions, where each solution represents a set of routes for the vehicles.
- The routes should ensure that all customers are visited exactly once and satisfy the capacity and time window constraints.

Step 2: Fitness evaluation

- Evaluate the fitness of each solution in the population.
- The fitness can be determined by the total distance traveled, considering any penalties for violating time window constraints or exceeding vehicle capacities.

Step 3: Selection

- Select parent solutions from the population for the crossover operation.
- The selection can be based on fitness values, employing techniques such as tournament selection or roulette wheel selection.

Step 4: Crossover

- Apply crossover operators, such as ordered crossover or partially mapped crossover, to create offspring solutions.
- The crossover should ensure that the offspring inherit good characteristics from the parent solutions.

Step 5: Mutation

- Apply mutation operators, such as swap mutation or insertion mutation, to introduce diversity in the population.
- The mutation helps explore new areas of the search space and avoid premature convergence.

Step 6: Fitness evaluation

- Evaluate the fitness of the offspring solutions.

Step 7: Replacement

- Select solutions from the population, including both parents and offspring, to form the new population.
- The replacement can follow strategies like elitism or generational replacement.

Step 8: Termination condition

- Check if the termination condition is met, such as reaching a maximum number of generations or a specific fitness threshold.
- If the condition is met, stop the algorithm; otherwise, go back to Step 2.

Step 9: Output the best solution

- Select the solution with the highest fitness from the final population as the best solution.
- This solution represents the optimized routes for the vehicles to deliver packages.

We give the algorithm implementation of the steps above:

3.2 Vehicle routing for waste collection

Waste collection is another important application of the VRP, where vehicles are assigned to collect waste from various locations efficiently. Let us consider a case where there are multiple waste collection points.

3.2.1 Problem formulation

Consider a scenario where waste collection vehicles need to be routed efficiently to collect waste from various locations. Each location has a specific waste quantity, and there are capacity constraints on the vehicles. The objective is to minimize the total distance traveled by vehicles while ensuring that the waste collection demand is met and the vehicle capacity is not exceeded.

Let:

- Set of waste collection points, $P = \{p_1, p_2, \dots, p_n\}$
- Set of vehicles $V = \{v_1, v_2, \dots, v_m\}$
- Location of each waste collection point $p \in P$ represented by coordinates $\left(x_p, y_p\right)$
- Waste quantity at each waste collection point$p \in P$ denoted by q_p
- Vehicle capacity for each vehicle $v \in V$ denoted by Q_v

Objective:

Minimize the total distance traveled by all vehicles while satisfying the waste collection demand and vehicle capacity constraints.

Let us denote the decision variables:

$x_{pvp} = 1$ if vehicle v visits waste collection point p, 0 otherwise.

d_{pq} = Euclidean distance between waste collection points p and q.

The mathematical formulation for the Vehicle Routing for Waste Collection problem can then be represented as:

Minimize:

$$\sum_{v \in V} \sum_{p \in P} \sum_{q \in P} \left(d_{pq} * x_{pvp} * x_{qv}\right) \tag{8}$$

Subject to:

1. $\sum_{v \in V} \sum_{p \in P} \sum_{q \in P} \left(q_p * x_{pvp}\right) \leq Q_v$, for all $v \in V$ (Vehicle capacity constraint)

2. $\sum_{v \in V} x_{pvp} = 1$, for all $p \in P$ (Each waste collection point is visited exactly once)

3. $\sum_{p \in P} x_{pvp} = 1$, for all $v \in V$ (Each vehicle visits exactly one waste collection point)

4. $x_{pvp} \in \{0, 1\}$, for all $p \in P, v \in V$ (Binary variable indicating whether a vehicle visits a waste collection point)

The objective function aims to minimize the total distance traveled by all vehicles. It sums up the distances between waste collection points p and q, multiplied by the binary variables x_{pvp} and x_{qv}, which indicate whether a vehicle visits those points.

The first constraint ensures that the waste collection demand at each waste collection point does not exceed the capacity of the vehicle assigned to it. It sums up the waste quantities q_p at each waste collection point p visited by vehicle v, multiplied by the binary variable x_{pvp}, and enforces that it must be less than or equal to the capacity Q_v of vehicle v.

The second constraint ensures that each waste collection point is visited exactly once. It states that the sum of binary variables x_{pvp} over all vehicles v must be equal to 1 for each waste collection point p.

The third constraint ensures that each vehicle visits exactly one waste collection point. It states that the sum of binary variables x_{pvp} over all waste collection points p must be equal to 1 for each vehicle v.

Finally, the binary variables x_{pvp} are constrained to be either 0 or 1, indicating whether a vehicle visits a waste collection point.

Solving this mathematical formulation using appropriate search algorithms can provide an optimized set of routes for the waste collection vehicles, minimizing the total distance traveled while meeting waste collection demand and vehicle capacity constraints. In this chapter, we give an example of using the Simulated Annealing algorithm to solve this.

3.2.2 Algorithm implementation: Ant colony optimization

Here's a step-by-step implementation of the Ant Colony Optimization (ACO) algorithm for solving the VRP for waste collection:

Step 1: Initialization

- Initialize the pheromone trails between waste collection points and depots.
- Set the parameters, such as the number of ants, maximum iterations, and pheromone evaporation rate.

Step 2: Ant solution construction:
Repeat the following steps for each ant:

- Choose a depot as the starting point.
- While there are unvisited waste collection points:
- Select the next waste collection point to visit based on the pheromone levels and heuristic information.
- Update the capacity of the ant's vehicle.
- Update the pheromone trail based on the visited waste collection point and the distance traveled.

Step 3: Update pheromone trails

- Update the pheromone trails based on the solutions constructed by the ants.
- The pheromone update can consider the quality of the solutions, such as the total distance traveled or the waste collection demand fulfilled.

Step 4: Termination condition

- Check if the termination condition is met, such as reaching a maximum number of iterations or a specific improvement threshold.
- If the condition is met, stop the algorithm; otherwise, go back to Step 2.

Step 5: Output the best solution

- Select the best solution found during the iterations.
- This solution represents the optimized routes for the waste collection vehicles.

3.3 Vehicle routing for emergency response

3.3.1 Problem formulation

Consider a scenario where emergency response vehicles need to be dispatched efficiently to emergency incidents with varying priorities and response time requirements. The objective is to minimize the total distance traveled by vehicles while considering the incident priorities and response time requirements.

Let:

- Set of emergency incidents $E = \{e_1, e_2, \dots, e_n\}$
- Set of vehicles $V = \{v_1, v_2, \dots, v_m\}$
- Location of each emergency incident $e \in E$ represented by coordinates (x_e, y_e)
- Priority of each emergency incident $e \in E$ denoted by p_e
- Response time requirement for each emergency incident $e \in E$ denoted by r_e
- capacity for each vehicle $v \in V$ denoted by Q_v.

Objective:
Minimize the total distance traveled by all vehicles while considering the incident priorities and response time requirements.
We shall now write the formulation mathematically.
Let us denote the decision variables:
$x_{eve} = 1$ if vehicle v is dispatched to emergency incident e, 0 otherwise.
d_{ef} = Euclidean distance between emergency incidents e and f.
The mathematical formulation for the Vehicle Routing for Waste Collection problem can be represented as follows:
Minimize:

$$\sum_{v \in V} \sum_{e \in E} \sum_{f \in E} (d_{ef} * x_{eve} * x_{fv}) \tag{9}$$

Subject to:

1. $\sum_{v \in V} \sum_{e \in E} (p_e * x_{eve}) \leq Q_v$, for all $v \in V$ (Vehicle capacity constraint)
2. $\sum_{v \in V} (x_{eve}) = 1$, for all $e \in E$ (Each emergency incident is assigned exactly one vehicle)
3. $\sum_{e \in E} (x_{eve}) \leq 1$, for all $v \in V$ (Each vehicle is assigned to at most one emergency incident)
4. $\sum_{v \in V} \sum_{e \in E} (r_e * x_{eve} \leq T_{max})$, for all $v \in V$ (Total response time of each vehicle must be within the maximum time limit)
5. $x_{eve} \in 0, 1$, for all $e \in E, v \in V$ (Binary variable indicating whether a vehicle is dispatched to an emergency incident)

The objective function aims to minimize the total distance traveled by all vehicles. It sums up the distances between emergency incidents e and f, multiplied by the binary variables x_{eve} and x_{fv}, which indicate whether a vehicle is dispatched to those incidents.

The first constraint ensures that the sum of incident priorities p_e at each emergency incident e dispatched to by vehicle v, multiplied by the binary variable x_{eve}, does not exceed the capacity Q_v of vehicle v.

The second constraint ensures that each emergency incident is assigned exactly one vehicle. It states that the sum of binary variables x_{eve} over all vehicles v must be equal to 1 for each emergency incident e.

The third constraint ensures that each vehicle is assigned to at most one emergency incident. It states that the sum of binary variables x_{eve} over all emergency incidents e must be less than or equal to 1 for each vehicle v.

The fourth constraint ensures that the total response time of each vehicle, considering the response time requirements r_e at each emergency incident, is within the maximum time limit T_{max}.

Finally, the binary variables x_{eve} are constrained to be either 0 or 1, indicating whether a vehicle is dispatched to an emergency incident or not.

By solving this mathematical formulation using appropriate search algorithms, an optimized set of dispatch routes for the emergency response vehicles can be obtained, minimizing the total distance traveled while considering incident priorities and response time requirements. Here, we give an algorithm implementation using Simulated Annealing.

3.3.2 Algorithm implementation: Simulated annealing

Here's a step-by-step implementation of the Simulated Annealing algorithm for solving the VRP for emergency response:

Step 1: Initialization

- Generate an initial solution, where each vehicle is assigned a set of emergency incidents to respond to.
- The solution should satisfy the capacity constraints and the response time requirements.

Step 2: Define objective function

- Define an objective function that combines the total distance traveled by vehicles with penalties for violating the response time requirements.

Step 3: Set initial temperature and cooling schedule

- Set an initial temperature for the simulated annealing algorithm.
- Define a cooling schedule to reduce the temperature over iterations.

Step 4: Iterate until termination condition.
While the termination condition is not met:

- Generate a neighboring solution by applying a neighborhood operator, such as swapping two incidents between vehicles or reassigning an incident to a different vehicle.
- Calculate the objective function difference between the current solution and the neighboring solution.

- Determine whether to accept the neighboring solution based on the objective function difference and the current temperature.
- Update the current solution if the neighboring solution is accepted.
- Update the temperature according to the cooling schedule.

Step 5: Output the best solution

- Select the best solution found during the iterations.
- This solution represents the optimized assignment of emergency incidents to vehicles.

3.4 Comparison of algorithm performance in different scenarios

To compare the performance of the search algorithms (Genetic Algorithm, Simulated Annealing, and Ant Colony Optimization) in different VRP scenarios, we can consider the following evaluation metrics:

I. **Objective function value:**

1. Calculate the objective function value (e.g., total distance traveled) for each algorithm in various problem instances.
2. Compare the objective function values to determine which algorithm yields better solutions.

II. **Computational time:**

1. Measure the computational time required for each algorithm to solve different problem instances.
2. Analyze the efficiency and scalability of the algorithms based on their computational time.

III. **Solution quality:**

1. Assess the quality of the solutions generated by each algorithm by comparing them to known optimal solutions or lower bounds.
2. Use metrics such as the percentage deviation from the optimal solution or the solution quality relative to a lower bound.

IV. **Robustness:**

1. Evaluate the robustness of each algorithm by considering how well they perform in the face of variations in problem parameters, such as the number of customers, vehicle capacities, or time window constraints.

2. Analyze the algorithms' ability to handle different problem instances with varying complexities.

V. **Sensitivity analysis:**

1. Conduct sensitivity analysis to examine the algorithms' sensitivity to changes in algorithm parameters, such as mutation rate or cooling schedule.

2. Identify the parameter settings that result in optimal performance for different scenarios.

VI. **Visualization:**

1. Visualize the routes and assignments generated by each algorithm to gain insights into their behavior and effectiveness.

2. Use graphical representations, such as route maps or heatmaps of pheromone trails, to compare the algorithm's solutions visually.

By comparing the algorithm performance in different scenarios using these evaluation metrics, we can assess their strengths, weaknesses, and suitability for specific VRP instances. The analysis will help in selecting the most appropriate algorithm based on the problem characteristics and requirements.

4. Factors influencing algorithm performance

In this section, we will discuss the key factors that influence the performance of search algorithms for solving the Vehicle Routing Problem (VRP). Understanding these factors is crucial for selecting the most appropriate algorithm and optimizing its performance.

4.1 Problem size and complexity

The size and complexity of the VRP instance play a significant role in determining the performance of search algorithms. The problem size refers to the number of customers or waste collection points, while complexity refers to the intricacy of the constraints and objectives involved.

As the problem size increases, the search space grows exponentially, making it more challenging to find an optimal solution within a reasonable amount of time. The complexity of the problem, such as the presence of time windows, vehicle capacities, or precedence constraints, further adds to the computational burden.

Search algorithms may struggle to find optimal solutions for large and complex VRP instances due to the increased search space and computational requirements. However, some algorithms, like genetic algorithms or ant colony optimization, are designed to handle larger problem sizes and complex constraints more effectively.

4.2 Algorithm parameters

The performance of search algorithms is influenced by various parameters that control their behavior and exploration-exploitation trade-off. It is essential to understand the impact of these parameters and tune them appropriately for different VRP instances.

Some common algorithm parameters include population size, mutation rate, crossover rate, pheromone evaporation rate, and cooling schedule. The values of these parameters can significantly impact the convergence speed, solution quality, and exploration capabilities of the algorithm.

For example, a higher mutation rate in genetic algorithms can promote exploration but may slow down convergence. On the other hand, a lower mutation rate can lead to premature convergence and suboptimal solutions. Similarly, the cooling schedule in simulated annealing determines the balance between exploration and exploitation.

Optimal parameter settings for search algorithms depend on the problem characteristics and the desired trade-offs between solution quality and computation time. Experimentation and parameter tuning techniques, such as grid search or evolutionary algorithms, can help identify suitable parameter values for specific VRP instances.

4.3 Problem-specific constraints

The VRP exhibits various constraints that must be considered during algorithm design and implementation. These constraints may include vehicle capacities, time windows, precedence relationships, or specific customer requirements.

The presence of constraints affects the search space and solution feasibility. For example, the vehicle capacity constraint ensures that the total demand of the assigned customers does not exceed the vehicle's capacity. The time window constraint requires deliveries to be made within specific time intervals.

Search algorithms need to incorporate these constraints in their formulations, fitness evaluation functions, and solution construction mechanisms. Violation of constraints often incurs penalties or leads to infeasible solutions. Therefore, algorithms must balance the exploration of the search space while ensuring constraint satisfaction.

Different algorithms may handle constraints differently. For example, genetic algorithms can use repair mechanisms to fix infeasible solutions, while ant colony optimization can enforce constraints during route construction by considering pheromone trails and heuristics.

Understanding the problem-specific constraints and selecting an algorithm that can handle them effectively are crucial for achieving feasible and optimized solutions in VRP.

4.4 Computational resources

The computational resources available, such as processing power, memory, and time, have a significant impact on algorithm performance. Search algorithms with high computational requirements may not be feasible to run on limited resources or within tight time constraints.

Large-scale and complex VRP instances may require significant computational resources to explore the vast search space and find optimal solutions. In such cases, parallel computing techniques or distributed algorithms can be employed to leverage multiple processors or computing nodes.

Consideration should also be given to the trade-off between solution quality and computational time. Some instances may require near-optimal solutions, while others may prioritize finding good-quality solutions within a reasonable time frame. To address the influence of computational resources on algorithm performance, several strategies can be employed:

I. **Algorithm optimization**

1. Enhance the efficiency of the algorithm by implementing algorithmic optimizations, such as pruning techniques, heuristic initialization, or local search operators.

2. Streamline the algorithm's operations to reduce computational overhead and improve convergence speed.

II. **Problem decomposition**

1. Decompose the problem into smaller subproblems that can be solved independently or in parallel.

2. Divide the VRP instance into clusters or zones and assign different computing resources to solve each subproblem concurrently.

3. Combine the solutions from the subproblems to obtain a solution for the entire problem.

III. **Approximation algorithms**

1. Utilize approximation algorithms that provide near-optimal solutions with lower computational requirements.

2. Approximation algorithms trade off optimality for computational efficiency and can be well suited for large-scale VRP instances.

IV. **Metaheuristic hybridization**

1. Combine different search algorithms or metaheuristics to leverage their strengths and compensate for their weaknesses.

2. Hybrid algorithms can take advantage of the exploration capabilities of one algorithm and the exploitation capabilities of another to improve overall performance.

V. **Resource scaling**

1. Scale the problem instance or adjust the level of detail to match the available computational resources.

2. For example, if the problem size is too large to solve optimally, consider aggregating or sampling the data to reduce the computational requirements while maintaining the problem's essential characteristics.

VI. **Hardware and parallel computing**

1. Utilize high-performance computing resources, such as multicore processors, distributed computing clusters, or cloud computing platforms, to accelerate algorithm execution.

2. Implement parallelization techniques, such as parallel algorithms, parallel fitness evaluation, or parallel solution construction, to exploit the available computing resources efficiently.

It is crucial to consider the computational resources when selecting an algorithm for VRP and to assess the algorithm's performance within the constraints of available resources. By optimizing algorithms, utilizing decomposition techniques, employing approximation methods, or leveraging parallel computing, it becomes possible to overcome limitations imposed by computational resources and find efficient solutions for VRP instances.

5. Recommendations for algorithm selection

Selecting the appropriate algorithm for solving the Vehicle Routing Problem (VRP) depends on the characteristics of the problem instance and the desired objectives. In this section, we provide recommendations for algorithm selection based on specific VRP scenarios.

5.1 Small-scale VRP instances

For small-scale VRP instances with a relatively small number of customers and simple constraints, exact algorithms, such as integer linear programming (ILP) formulations or branch-and-bound methods, can be effective. These algorithms aim to find the globally optimal solution by exploring the entire search space.

5.1.1 ILP formulation

1. Formulate the VRP as an ILP problem using binary decision variables to represent the assignment of customers to vehicles.

2. Define objective function and constraints, including vehicle capacity, time windows, and precedence relationships.

3. Use ILP solvers, such as CPLEX or Gurobi, to solve the formulation and obtain an optimal solution.

5.1.2 Branch-and-bound method

1. Employ a tree-based search approach to explore the search space systematically.

2. Divide the problem into subproblems and use lower bounds to prune infeasible or non-promising branches.

3. Continually refine the lower bounds and update the best solution found during the search process.

5.2 Large-scale VRP instances

For large-scale VRP instances with a high number of customers and complex constraints, metaheuristic algorithms are more suitable due to their ability to handle larger search spaces efficiently.

Genetic algorithms, ant colony optimization, and particle swarm optimization are popular metaheuristic algorithms for large-scale VRP instances. These algorithms can explore the search space effectively and provide near-optimal solutions within a reasonable computational time.

In addition, decomposition techniques, such as clustering or partitioning the problem into smaller subproblems, can be applied to further enhance the scalability of the algorithms.

5.3 VRP instances with time windows

When dealing with VRP instances that have time window constraints, algorithms that can effectively handle time-dependent constraints are recommended.

Simulated annealing, tabu search, and variable neighborhood search algorithms are well suited for VRP instances with time windows. These algorithms can incorporate time constraints into their formulations and neighborhood structures to ensure that feasible solutions are generated.

Incorporating time window penalties or soft time window constraints can also be useful to handle situations where strict adherence to time windows is challenging.

5.4 VRP instances with multiple objectives

For VRP instances with multiple objectives, such as minimizing the total distance traveled and maximizing customer satisfaction simultaneously, multi-objective metaheuristic algorithms are appropriate.

Algorithms like the NSGA-II (Non-dominated Sorting Genetic Algorithm II) [14] or SPEA2 (Strength Pareto Evolutionary Algorithm 2) [15, 16] can handle multiple objectives and generate a set of solutions that represent the trade-off between conflicting objectives. These algorithms use dominance-based sorting and selection mechanisms to maintain a diverse set of non-dominated solutions, known as the Pareto front. Decision-makers can then choose the solution from the Pareto front that best aligns with their preferences.

Additionally, the use of preference-based approaches or weighting schemes can help find a single solution that combines multiple objectives into a single objective function [17].

It is important to note that the recommendations provided are general guidelines, and the selection of the most appropriate algorithm for a specific VRP instance should be based on thorough experimentation, considering the problem characteristics, available computational resources, and desired trade-offs between solution quality and computation time.

6. Conclusion

In this chapter, we have explored the application of search algorithms for solving the Vehicle Routing Problem (VRP). We provided an overview of the VRP and its variants, discussing the main challenges and constraints associated with the problem. We then conducted a comparative analysis of various search algorithms used to solve the VRP, including genetic algorithms, simulated annealing, ant colony optimization, and other metaheuristic approaches.

Through our analysis, we found that different search algorithms have their strengths and weaknesses when applied to the VRP. Genetic algorithms excel in handling large-scale instances and complex constraints, while simulated annealing is effective for problems with time window constraints. Ant colony optimization demonstrates robustness in finding good-quality solutions, and other algorithms like tabu search, particle swarm optimization, and iterated local search offer alternative approaches to tackle the VRP.

In the case study section, we provided examples of how search algorithms have been applied to vehicle routing for package delivery, waste collection, and emergency response. We compared the algorithm performance in different scenarios, considering factors such as solution quality, computation time, and constraint satisfaction.

The findings of this chapter highlight the importance of considering problem-specific characteristics, such as problem size, complexity, and constraints, when selecting an algorithm for the VRP. The computational resources available also play a significant role in algorithm performance, and different strategies can be employed to optimize performance within resource constraints.

6.1 Summary of findings

Based on our analysis and case studies, we observed that no single algorithm performs optimally for all types of VRP instances. The choice of algorithm depends on the problem characteristics, including the problem size, complexity, time windows, and multiple objectives. It is essential to evaluate and compare the performance of different algorithms using appropriate metrics to determine the most suitable algorithm for a given VRP instance.

Metaheuristic algorithms, such as genetic algorithms, simulated annealing, and ant colony optimization, demonstrate their effectiveness in solving the VRP. These algorithms provide a good balance between exploration and exploitation, enabling them to handle the complex search space and constraints associated with the VRP.

Additionally, hybridization techniques, parallel computing, and problem decomposition can further enhance the algorithm performance and scalability for large-scale VRP instances.

6.2 Contributions of the study

This chapter contributes to the understanding and analysis of search algorithms for solving the VRP. The key contributions of this study are as follows:

1. Comprehensive overview: We provided a comprehensive overview of the VRP, its variants, and the main challenges and constraints associated with the problem.

2. Comparative analysis: We conducted a comparative analysis of different search algorithms, including genetic algorithms, simulated annealing, ant colony optimization, and other metaheuristic approaches, highlighting their strengths and limitations.

3. Case studies and examples: We presented case studies and examples of how search algorithms have been applied to real-world scenarios, such as vehicle routing for package delivery, waste collection, and emergency response.

4. Factors influencing algorithm performance: We discussed the factors that influence the performance of search algorithms, including problem size, complexity, algorithm parameters, problem-specific constraints, and computational resources.

6.3 Future research directions

While this chapter provides a comprehensive analysis of search algorithms for the VRP, there are several avenues for future research that can further enhance the field:

1. Development of hybrid algorithms: Investigate the development of hybrid algorithms that combine the strengths of multiple algorithms to improve solution quality and convergence speed for the VRP.

2. Incorporation of uncertainty: Explore the integration of uncertainty into the VRP models and algorithms, considering factors such as traffic congestion, uncertain customer demands, and dynamic time windows.

3. Integration of machine learning: Investigate the integration of machine learning techniques, such as deep learning or reinforcement learning, into search algorithms for the VRP. Machine learning can be used to enhance the exploration and exploitation capabilities of algorithms, improve decision-making processes, and adapt the algorithms to evolving problem conditions.

4. Metaheuristic parameter tuning: Conduct further research on parameter tuning techniques for metaheuristic algorithms to automatically adjust algorithm parameters based on problem characteristics and instance-specific requirements. This can improve the algorithm's adaptability and performance across different VRP instances.

5. Multi-objective VRP: Explore advanced techniques for solving VRP instances with multiple objectives, such as developing new algorithms or improving existing algorithms, to better handle the trade-offs between conflicting objectives. This can include the development of preference-based approaches, interactive algorithms, or Pareto dominance-based methods.

6. Real-time VRP: Investigate real-time or dynamic VRP scenarios where new customer requests arrive during the execution of the routing plan. Develop algorithms that can efficiently adapt the existing routes and schedules to incorporate the dynamic changes in real time, considering constraints such as time windows, vehicle capacities, and customer priorities.

7. Benchmark problems and performance metrics: Continuously improve the benchmark problems and performance metrics used for evaluating VRP algorithms. Develop more realistic and challenging problem instances that better capture the complexities of real-world scenarios, and define performance metrics that encompass a wider range of evaluation criteria, including efficiency, robustness, and sustainability.

8. Collaboration and cooperative algorithms: Explore the application of collaboration or cooperative algorithms for the VRP, where multiple vehicles or agents work together to optimize the routing and scheduling process. Investigate communication protocols, coordination mechanisms, and distributed algorithms to solve complex VRP instances more efficiently.

By pursuing these future research directions, we can further advance the field of VRP and develop more effective algorithms to address the challenges and complexities encountered in real-world vehicle routing applications. These advancements will contribute to improving the efficiency, sustainability, and operational effectiveness of transportation and logistics systems.

Author details

Oladimeji Samuel Sowole
African Institute for Mathematical Sciences, Senegal

*Address all correspondence to: oladimeji.s.sowole@aims-senegal.org

References

[1] Golden BL, Raghavan S, Wasil EA. The Vehicle Routing Problem: Latest Advances and New Challenges. Vol. 43. Springer Science & Business Media; 2008

[2] Laporte G. The vehicle routing problem: An overview of exact and approximate algorithms. European Journal of Operational Research. 1992;**59** (3):345358

[3] Semet F, Toth P, Vigo D. Chapter 2: Classical exact algorithms for the capacitated vehicle routing problem. In: Vehicle Routing: Problems, Methods, and Applications. Second ed. SIAM; 2014. pp. 37-57

[4] Kallehauge B, Larsen J, Madsen OBG, Solomon MM. Vehicle Routing Problem with Time Windows, Column Generation. Springer; 2005. pp. 67-98

[5] Alba E, Dorronsoro B. Solving the vehicle routing problem by using cellular genetic algorithms. In: Lecture Notes in Computer Science, Proceedings of the Conference on Evolutionary Computation in Combinatorial Optimization, LNCS, Coimbra, Portugal, 5–7 April 2004. Vol. 3004. Berlin/Heidelberg, Germany: Springer; 2004. pp. 11-20

[6] Vaira G. Genetic Algorithm for Vehicle Routing Problem. Ph.D. Thesis. Vilnius, Lithuania: Vilnius University; 2014

[7] Yu VF, Susanto H, Jodiawan P, Ho T-W, Lin S-W, Huang Y-T. A simulated annealing algorithm for the vehicle routing problem with parcel lockers. IEEE Access. 2022;**10**:20764-20782. DOI: 10.1109/ACCESS.2022.3152062

[8] Wang C, Zhao F, Mu D, Sutherland JW. Simulated annealing for a vehicle routing problem with simultaneous pickup-delivery and time windows. In: Prabhu V, Taisch M, Kiritsis D, editors. Advances in Production Management Systems. Sustainable Production and Service Supply Chains. APMS 2013. IFIP Advances in Information and Communication Technology. Vol. 415. Berlin, Heidelberg: Springer; 2013. DOI: 10.1007/978-3-642-41263-9_21

[9] Calvete HI, Galé C, Oliveros MJ. Ant Colony optimization for solving the vehicle routing problem with delivery preferences. In: Engemann KJ, Gil-Lafuente AM, Merigó JM, editors. Modeling and Simulation in Engineering, Economics, and Management. MS 2012. Lecture Notes in Business Information Processing. Vol. 115. Berlin, Heidelberg: Springer; 2012. DOI: 10.1007/978-3-642-30433-0_23

[10] Gurpreetsingh E, Dhir V. Vehicle Routing Problem by Ant Colony Optimization. Semantic Scholar; 2014

[11] Barbarosoglu G, Ozgur D. A tabu search algorithm for the vehicle routing problem. Computers & Operations Research. 1999;**26**(3):255-270. DOI: 10.1016/S0305-0548(98)00047-1

[12] Marinakis Y, Marinaki M, Migdalas A. Particle swarm optimization for the vehicle routing problem: A survey and a comparative analysis. In: Martí R, Pardalos P, Resende M, editors. Handbook of Heuristics. Cham: Springer; 2018. DOI: 10.1007/978-3-319-07124-4_42

[13] Euchi J, Zidi S, Laouamer L. A hybrid approach to solve the vehicle routing problem with time windows and synchronized visits In-home health care. Arabian Journal for Science and Engineering. 2020;**45**:10637-10652. DOI: 10.1007/s13369-020-04828-5

[14] Deb K, Pratap A, Agarwal S, Meyarivan T. A fast and elitist multiobjective genetic algorithm: NSGA-II. IEEE Transactions on Evolutionary Computation. 2002;**6**(2): 182-197. DOI: 10.1109/4235.996017

[15] Amuso VJ, Enslin J. The strength of Pareto evolutionary algorithm 2 (SPEA2) applied to simultaneous multi-mission waveform design. In: 2007 International Waveform Diversity and Design Conference, Pisa, Italy. Vol. 20. 2007. pp. 407-417. DOI: 10.1109/21WDDC.2007.4339452

[16] Long J, Sun Z, Pardalos PM, Hong Y, Zhang S, Li C. A hybrid multi-objective genetic local search algorithm for the prize-collecting vehicle routing problem. Information Sciences. 2019;**478**:40-61. DOI: 10.1016/j.ins.2018.11.006

[17] Ait Haddadene SR, Labadie N, Prodhon C. Bicriteria, vehicle routing problem with preferences and timing constraints in home health care services. Algorithms. 2019;**12**:152. DOI: 10.3390/a12080152

Chapter 12

An Overview of Ant Colony Optimization Algorithms for Dynamic Optimization Problems

Alireza Rezvanian, S. Mehdi Vahidipour and Ali Sadollah

Abstract

Swarm intelligence is a relatively recent approach for solving optimization problems that usually adopts the social behavior of birds and animals. The most popular class of swarm intelligence is ant colony optimization (ACO), which simulates the behavior of ants in seeking and moving food. This chapter aim to briefly overview the important role of ant colony optimization methods in solving optimization problems in time-varying and dynamic environments. To this end, we describe concisely the dynamic optimization problems, challenges, methods, benchmarks, measures, and a brief review of methodologies designed using the ACO and its variants. Finally, a short bibliometric analysis is given for the ACO and its variants for solving dynamic optimization problems.

Keywords: ant colony optimization, dynamic optimization problem, metaheuristics, swarm intelligence, dynamic environments

1. Introduction

Swarm intelligence is a relatively new approach to optimizing problems that typically mimics the social behavior of birds and animals. Ant colony optimization (ACO), which models ant behavior in locating and moving food, is the most prominent type of Swarm intelligence. The early ACO algorithms were applied to discrete optimization problems such as traveling salesman problems, routing, and scheduling. These types of problems consist of a limited number of fixed parameters in stationary environments. However, most real problems have dynamic and time-varying parameters with a continuous range of values. It means the optimal solution is not fixed to changing environments and time. Therefore, the optimization algorithm needs a strategy for tracking optimal solutions. In addition to solving static optimization problems, the ACO algorithm and its variants can also solve dynamic optimization problems.

The rest of this chapter is organized as follows. Section 2 describes the preliminary and background of the work, including the ACO and dynamic optimization problems. Section 3 briefly overviews evolutionary algorithms and ACO for dynamic optimization problems. Section 4 gives a bibliometric analysis of the ACO and its variants for dynamic optimization problems. Finally, Section 5 concludes this chapter.

IntechOpen

2. Preliminary and background

In this section, the ACO and dynamic optimization problems are described briefly.

2.1 Ant colony optimization

The Ant Colony Optimization (ACO) algorithm successfully solves various discrete optimization problems. The standard algorithm of ACO was proposed by Dorigo as a multi-agent approach for solving the traveling salesman problem (TSP) [1]. The main idea of the ACO algorithm is inspired by the behavior of colonies of ants seeking food. Usually, at first, ants search their environment randomly for food and return some of the food to their nest once found. They also leave a pheromone on the path they found. The value of pheromones left on their path depends on the quality and size of the food source, which evaporates gradually. The remaining pheromones on a path can persuade other ants to follow the path. After a short time, most ants can follow the shorter path specified by the strongest pheromone. A general pseudocode of the ACO algorithm is given in Algorithm 1.

Algorithm 1. Pseudocode of standard ACO for solving optimization problems.

```
Ant Colony Optimization Algorithm
Begin
    Initialize user parameters of ACO and generate the initial population of the ant colony
While the stopping condition is not met do
        Position each ant in a starting node
Repeat
For each do
            Choose the next node by applying the state transition rule
            Apply step-by-step pheromone update
End for
 Until every ant has built a solution
        Update the best solution
        Apply offline pheromone update
 End While
End
```

Through the running of the algorithm, ants first produce different solutions randomly. Then, the solutions are improved by updating the pheromones according to the type of the problem and graph traversal, pheromones set on vertices or edges of the graph. Traversing the edge between two nodes i and j depends on the probability of edge, which is calculated as given follows:

$$p_{ij}^{k} = \frac{\tau(t)_{ij}^{\alpha}\eta_{ij}^{\beta}}{\sum_{j\epsilon N^{k}(i)} \tau(t)_{ij}^{\alpha}\eta_{ij}^{\beta}} \tag{1}$$

where, p_{ij}^{k} is the probability of traversing from the edge between nodes i and j, $\tau(t)_{ij}^{\alpha}$ denotes the value of the pheromone trail, η_{ij}^{β} is the heuristic information, and $N^{k}(i)$ is the set of nodes that can be visited by ant k. Before updating the pheromones, it is

DOI: http://dx.doi.org/10.5772/intechopen.111839

recommended to perform a local search to improve results. However, the method of updating the pheromones is suggested as given follows [1]:

$$\tau_{ij}(t+1) = (1-\rho).\tau_{ij}(t) + \Delta\tau_{ij}(t) \tag{2}$$

where ρ is the evaporation rate of pheromone and $\Delta\tau_{ij}(t)$ denotes the amount of pheromone trail put on the edge between nodes i and j. That amount is calculated according to the quality of the solutions found by the whole colony, which include the edge between nodes i and j.

2.2 Dynamic optimization problems

A wide range of real-world optimization problems is dynamic, in which the optimal solution(s) to the problems may change over time. An example of such problems includes the Dynamic Vehicle Routing Problem (DVRP) [2], where the online arrival of customers' requests during operation is imposed as the source of dynamism in static vehicle routing, whereby previously found solution(s) may no longer be valid. Another well-known example is the Dynamic Shortest Path Routing Problem (DSPRP) [3], where optimization aims to find the shortest path from a specific source to a specific destination in a changing network environment while minimizing the total cost associated with the path.

As a formal definition [4], a dynamic optimization problem can be defined by a quintuple $\{\Omega, x, \o, f, t\}$, where Ω is the search space, x denotes a feasible solution in Ω, $\o$ is the system control parameters which determine the solution distribution in the fitness landscape, f represents the static objective function and t denotes the time. To define a dynamic environment, different change types can be defined. Yang and Pelta [5] suggested a framework of eight change types: small step change, large step change, random change, chaotic change, recurrent change, recurrent change with noise, and dimensional change. The purpose of optimization in such problems is no longer just locating the optimal solution(s) but instead tracking the shifting optima over time.

In general, all aspects of science and engineering include optimizing a set of complex problems. In these cases, the optimization objectives, some restrictions, or other elements of the problems may vary over time. In these cases, the optimum solution(s) to the problem are also subject to change. As seen from the examples mentioned earlier, unlike static optimization problems, the goal of optimization in dynamic problems is no longer locating the optimal solution(s) but tracking the trajectories of the optimum(s) over time with a high level of accuracy.

2.2.1 Dynamic optimization problem categories

Weicker et al. [6] proposed a framework for dynamic optimization problems based on coordinate transformations (i.e., change in the position of the optima) and fitness rescaling (i.e., change in the value of the objective function of the optima). Eberhart et al. [7] classified dynamic environments into three major types: Type I, Type II, and Type III. In type I environments, the position of the optima changes over time; however, its objective function's value remains unchanged. In type II, the position of the optima does not change but its value varies over time. Finally, type III, where both the position and the value of the optima change over time.

Angeline [8] classified different dynamic environments based on the trajectories of peaks into three categories: linear, circular, and random. According to Duhain et al. [9], dynamic environments can be categorized based on how severe the changes are over time and space. They call them behavioral classes, including progressively changing, abruptly changing, and chaotic environments. Progressively changing environments have frequent temporal changes but small spatial changes, and they change gradually and smoothly. In abruptly changing environments, the changes happen rarely, but the spatial alterations are severe. Finally, if both the spatial and temporal severity of the dynamic environment is high, the behavior of the environment is chaotic. Moreover, they combined their behavioral classification with Eberhart et al. [7] and Angeline's [8] types and defined 27 kinds of dynamic environments based on the direction, the trajectory, and the severity of the changes that the environments undergo. A more comprehensive study of dynamic optimization problems with a wider scope can be found in the literature [10–15].

2.2.2 Optimization challenges in dynamic optimization problems

The purpose of optimization in dynamic optimization problems is no longer to locate the stationary optimal solution(s) but to track the trajectories of the optimum (s) over time as accurately as possible [16–18]. This situation possesses additional challenges that should be addressed properly to obtain promising results. Reacting to changes plays a critical role in maintaining the performance of optimization algorithms in dynamic environments. The first step toward reacting to changes is to detect them in the first place and then adopt a suitable strategy to deal with them.

Richter [19] discussed two major types of change detection, referred to as population-based detection and sensor-based detection. In population-based detection, statistical hypothesis testing is performed at each generation to see whether the alteration in the fitness of the individuals was not due to their convergence. In sensor-based detection, some measurements, so-called "fitness landscape sensors," are placed either randomly or on a regular pattern into the landscape. At every generation, the environment changes if any of the sensors detect an altered fitness value. As mentioned earlier, the two mechanisms have their own merits and limits. Population-based detection does not impose any additional fitness function evaluations on the algorithms, while sensor-based detection, on the other hand, can eliminate elaborate statistical hypothesis testing procedures. Many existing works in the literature reevaluate the previous best solutions to detect environmental changes [20] ((References)). This strategy can be regarded as a combination of both mentioned change detection types, as reevaluating previous best solutions can be considered a kind of sensor-based detection where the previous best individuals dynamically allocate the distribution of the sensors in the landscape.

For optimization algorithms to hopefully advance the search process in dynamic environments, two issues must be addressed properly. The first issue arising from the change in the environment is outdated information, causing the quality of previous solutions to be compromised. This has a deleterious effect on the performance of the optimization and can mislead the search process toward the wrong positions in the landscape. This problem can be eliminated by reevaluating past information or replacing it with current information. The second issue is diversity loss, which is much more serious than the other challenges. This issue arises when the population of a population-based heuristic converges to a peak. When the environment changes, if the peak goes out of the population's spatial range, the function evaluations required for a partially converged population to re-diversify and re-converge to the shifted peak are quite detrimental to the performance [21]. Therefore, maintaining diversity is essential for population-based methods to adapt the imposed changes.

2.2.3 Dynamic optimization problems benchmarks

Recently, there has been a growing interest in the resolution of synthetic dynamic optimization problems due to the ability of these models to approximate a variety of real-world situations closely. The first and most widely used real-valued synthetic dynamic optimization test suite in the literature is the Moving Peaks Benchmark (MPB) proposed by Branke [22], which is highly regarded due to its configurability. The baseline function for MPB is specified as given in the following equation:

$$\mathrm{f_t(x)} = \max_{i\in\{1,\ldots,m\}} \frac{H_t(i)}{1 + W_t(i)\sum_{j=1}^{D}(x_t(j) - X_t(i,j))^2}, \tag{3}$$

where m is the number of peaks, $x_t(j)$ is the jth parameter of solution x in a d-dimensional search space, $X_t(i,j)$ is the jth parameter of the center of ith peak in the tth time, $H_t(i)$ is the height of ith peak, and $W_t(i)$ is the width of the ith peak the time t.

The Gaussian Peak Bench (GPB) [23] provides another method for generating moving peaks. In the Generalized moving peaks benchmark (GMPB) [24], the baseline function generates landscapes created by collecting some components. The generated modules by GMPBs can differ from symmetric to asymmetric, unimodal to multimodal, flat to highly irregular, and circular contour lines to highly ill-conditioned.

2.2.4 Evaluation measures

Several measurement metrics are proposed to evaluate the optimization algorithms' performance and different aspects of the methods for dynamic optimization problems. For instance, Weicker [25] introduced accuracy, stability, and reactivity as key optimization characteristics in dynamic environments. Alba et al. [26] proposed $\beta_{\text{degradation}}$ to measure the degradation of optimization algorithms in dynamic environments.

A more general approach to measuring an attribute of dynamic optimization problems is proposed in Ref. [27], which calculates an attribute of a population as the area below the curve. Ayvaz et al. [28] proposed a dissimilarity factor, which evaluates the ability of an optimization algorithm to recover after a change. Recently, Kordestani et al. [29] proposed two measures to compare algorithms in dynamic environments. The first measure is alpha-accuracy, which calculates the degree of closeness of the optimization error to a certain expected level during the optimization process. The second measure is fitness adaptation speed which evaluates the average speed of algorithms in adapting to environmental changes. Despite the number of performance measures in the literature, most authors have used offline error for measuring their proposed algorithms. This is because they can compare their methods with other peer algorithms. This measure is defined as given follows [17, 29–31]:

$$E_O = \frac{1}{T}\sum_{\mathrm{t}=1}^{T} e_t^* \tag{4}$$

Name	References	Measure
Accuracy	[25, 27]	The goodness of the best-found solution in the current population
Stability	[25, 27]	The impact of changes in the environment on the optimization accuracy
Reactivity	[25, 27]	The ability of the algorithm to react quickly to changes
Offline Error	[35]	The average fitness error of the best-found solution at every point in time
Best Error Before Change	[32, 34, 36]	The average fitness error of the best-found solution over the environmental changes
Fitness Degradation	[26]	The degradation in the quality of solutions as changes in the landscape occur
Area Below Curve	[27, 37]	An attribute of the population is the area below the curve
Dissimilarity Factor	[28]	The ability of an optimization algorithm to recover after a change
Alpha-Accuracy	[20, 29]	The degree of closeness of the optimization error to a certain expected level during the optimization process
Fitness Adaptation Speed	[20, 29]	The average speed of the algorithms in adapting to the environmental changes

Table 1.
Summary of the performance measures for dynamic optimization problems.

where T is the maximum number of function evaluations and e_t^* is the minimum error gained by the optimization algorithm at the tth fitness evaluation. Another measure used in Ref. [32] has also been termed offline error as the best error before the change. This measure is slightly different from the previously mentioned measure in Eq. (4) and may become a source of confusion when comparing different methods. This measure, first proposed in Ref. [33] as an accuracy criterion and then named the best error before change by Nguyen et al. [12], is calculated as the average of the minimum fitness error achieved by the algorithm at the end of each period right before the moment of change, as given follows [34]:

$$E_B = \frac{1}{K}\sum_{k=1}^{K}(h_k - f_k) \tag{5}$$

where f_k is the best solution obtained by an algorithm just before the kth change happens, h_k is the optimum value of the kth environment, and K is the total number of environments—the only difference between Eqs. (4) and (5) are the time taken to record the error gained by the algorithm, where in Eq. (4), the error is recorded every time the function is evaluated, and in Eq. (5), it is recorded right before the change. **Table 1** summarizes some of the existing performance measures in the literature.

3. Related works

3.1 Evolutionary computation for dynamic optimization problems

Since exact algorithms are impractical in dynamic environments, stochastic optimization techniques have become popular. Among them, evolutionary computation

(EC) techniques have attracted much attention due to their potential for solving complex optimization problems. Nevertheless, evolutionary computation methods should undergo certain adjustments to work well when applied to dynamic optimization problems. Diversity loss is the most severe challenge to evolutionary computation methods in dynamic optimization problems. This issue arises from the tendency of individuals to converge on a single optimum. As a result, when the global optimum is shifted away, the number of function evaluations (FEs) required for a partially converged population to relocate the optimum is quite harmful to performance.

Over the years, researchers have proposed various techniques to improve traditional evolutionary computation methods for solving dynamic optimization problems. According to Reference Nguyena et al. [12], the existing proposals can be categorized into the following six approaches:

1. Increasing the diversity after detecting a change in the environment

2. Maintaining diversity during the optimization process

3. Employing memory schemes to retrieve information about previously found solutions

4. Predicting the location of the next optimal solution(s) after a change is detected

5. Making use of the self-adaptive mechanisms of evolutionary computations

6. Using multiple sub-populations to handle separate areas of the search space concurrently

As mentioned, the multi-population (MP) approach effectively handles dynamic optimization problems, especially in multimodal fitness landscapes. The success of MP methods can be attributed to the following three reasons [12]:

1. Population diversity can still be maintained globally even when populations search within different subareas of the fitness landscape.

2. It is possible to locate and track multiple changing optima simultaneously. This feature can facilitate tracking of the global optimum, given that one of the being tracked local optima may become the new global optimum when environmental changes occur.

3. Extending any single-population approach, e.g., diversity increasing/maintenance schemes, memory schemes, and adaptive schemes, is easy.

There are several studies in the literature on dynamic optimization problems in general. For instance, Jin and Branke [13] addressed and discussed various types of uncertainty in evolutionary optimization. Cruz et al. [11] have contributed to the advancement of knowledge and relevance of the field by accomplishing a twofold task: (1) They have established a public web repository of a large number of relevant references about the topic over the past decade and then categorized the references based on the type of publications, type of dynamism, methods for solving dynamic optimization problems, performance measures, applications and publication year;

(2) Afterward, they have performed an overview of the research carried out on dynamic optimization problems based on the collected repository. Nguyen et al. [12] have provided an in-depth survey of evolutionary optimization in dynamic environments. The study examined state-of-the-art academic research from four different points of view: benchmark problems, performance measures, methodology, and theory.

Yazdani et al. [14, 15] in two parts of survey papers, represented a review of research studies regarding single-objective unconstrained continuous dynamic optimization problems. Since an efficient dynamic optimization algorithm consists of several parts to cope with dynamic optimization problems, they tried to provide a comprehensive taxonomy to identify the parts of dynamic optimization algorithms. In the second part of this survey, they gave an overview of dynamic optimization problem benchmarks and performance measures.

Another study in the literature is provided by Moser et al. [38]. In their work, the authors have given an overview of the available approaches in the literature for solving MPB. They have examined various methods for the MPB in four groups: Evolutionary Algorithms, Swarm Intelligence Algorithms, Hybrid Approaches, and Other Approaches. Although their work is very valuable in studying and summarizing different methods for solving the MPB, it does not provide categorical information on how different methods work and which mechanisms can improve the performance of different methods. In addition, it does not explain the reasons for the superiority of specific approaches. Therefore, we believe it is necessary to provide a well-structured survey of available methods for solving the MPB to extract their strengths and weaknesses and open the field to develop new efficient approaches to address dynamic optimization problems.

3.2 ACO for dynamic optimization problem

Ant colony optimization is a candidate suited for solving dynamic optimization problems due to its ability to adapt to changing environments. It searches for optimal solutions in time-varying environments. Basically, this is accomplished through the use of pheromones. These are chemical signals deposited by ants as they move through the environment. Also, because of the evaporating nature of pheromones, ants can quickly find promising regions of the solution space. They can converge on proper solutions even as the problem changes over time.

In the following, brief overviews of the studies regarding ACO methods for the dynamic optimization problem are presented. Dréo and Siarry [39] is the first study of ant colonies for optimization in a continuous and dynamic environment. The Dynamic Hybrid Continuous Interacting Ant Colony (DHCIAC), as a hierarchical algorithm focused on the communication behaviors of the ants and hybridized an interacting ACO for collecting information about the environment with a Nelder-Mead local search [40] for improving the solutions. The interacting algorithm was originally designed for a static environment that used several communication channels with different characteristics. The first channel is stigmergic, an indirect coordination mechanism between agents and their environments. Ants move around pheromone regions due to their role. The second channel is the direct channel where ants can communicate through messages. Each ant has a stack of messages and can send messages to the other ants. Also, based on a parameter, the ants can follow a new solution or boost the current one. According to the hierarchical framework of the algorithm and its channels, the DHCIAC can propagate information regarding the changing environment. Finally, an efficient local search can improve results. The authors also developed two versions of DHCIAC with different mechanisms for finding dynamic optimal. DHCIAC_track uses an enhanced local search to track the optimal evolution, while DHCIAC_find uses a classical Nelder-Mead local search to find the optimal solution whenever the environment changes.

The charged ants for continuous dynamic optimization are represented by Tfaili et al. [41]. The authors proposed assigning a repulsive electric charge to each ant. This charge prevents ants from getting close to maintaining a certain level of diversity. In this algorithm, the ants initially distribute in the environment randomly. Then, the new continuous solutions are generated based on a weighted Gaussian distribution as pheromones. The charged values of ants can be positive or negative based on the quality of the best solutions. The algorithm works with a limited and fixed population, and each ant can produce a solution similar to ACO [42]. As each interaction proceeds, the better new solutions replace the older and worse solutions.

As an application of the ant colony optimization, Montemanni et al. [43] presented the dynamic version of the ACO for the Vehicle Routing Problem (VRP), where the problem parameters change over time. Dynamic VRP aims to find an optimal route plan for a fleet of vehicles. This route plan considers changes in parameters such as customer demands, vehicle availability, and traffic conditions. Three modifications are incorporated into the ACO algorithm to cope with this problem, including local search, ant behavior, and pheromone updating. With local search, the quality of the solutions can be improved as the parameters of the problem change over time. Ant behavior modification can improve ants' behavior to adjust the changing parameters. Finally, by pheromone updating modification, pheromone updating adjusted the parameters over time.

Korosec et al. suggested a differential ant-Stigmergy algorithm for dynamic optimization problems [44]. Stigmergy is commonly used in multi-agent systems as a kind of indirect cooperation or coordination between agents and the environment. An optimization process boosts the best current solution by creating a proper path. The path is based on graph search using pheromones assigned to a Cauchy distribution. Then a path from a source node to a destination node is formed based on choosing each ant from another ant according to its probability. This process is repeated until reaching a destination node. By finding paths, the paths are evaluated, and the pheromones are updated.

In [45], the authors introduced a dynamic strategy to adapt the parameters of ACO through the search process dynamically. The strategy helps the algorithm adapt to environmental changes and prevent falling into local optima. Besides, they utilize a mechanism based on entropy to balance exploration and exploitation in the search process to improve the quality of solutions. Finally, the algorithm is applied to solving the TSP and VRP.

In [46], Zhou et al. represented an improved version of ACO, which incorporates two significant mechanisms parameter adaptation and dynamic hybridization. The parameter adaptation mechanism encompasses adapting the parameters of the ACO algorithm dynamically based on the features of the problem being solved. This mechanism better guides the algorithm to balance the search process's exploration and exploitation. The dynamic hybridization mechanism hybridizes the ACO algorithm with another metaheuristic algorithm, such as Genetic Algorithm or Simulated Annealing, depending on the growth of the

Reference(s)	Problem
[49]	Dynamic Electric Vehicle Dispatch Optimization
[50]	Dynamic Chemical Processes
[51]	Dynamic Cold Chain Logistics Scheduling
[52]	Dynamic Distribution Network
[53]	Dynamic Knapsack Problem
[54, 55]	Dynamic Load Balancing
[56]	Dynamic Location Routing Problem
[48]	Dynamic Quadratic Assignment Problem
[57]	Dynamic Railway Junction Rescheduling Problem
[58]	Dynamic Traveling Salesman Problem
[59, 60]	Dynamic Vehicle Routing Problem

Table 2.
Summary of ACO applications for dynamic optimization problems.

search process. This idea allowed the algorithm to avoid local optima and keep the diversity of solutions.

Brest et al. [47] introduced two population-based algorithms based on differential evolution and ACO with continuous variables: the self-adaptive differential evolution algorithm (jDE) and the differential ant-stigmergy algorithm (DASA). These algorithms use the behavior of ants to guide their search process, and ants can communicate using pheromones in the environment to identify the optimal solution. They focused on comparing the performances of these two algorithms.

In [48], Guntsch et al. proposed a population-based ant colony optimization (PBACO) as an improved version of the ACO algorithm that employs several colonies to search for solutions. In the PBACO, a mechanism is designed for handling changes in the problem environment over time. Also, an algorithm is devised with multi-population in which each population is responsible for searching for solutions in a different time of the environment. Then, the best solution for each population is shared among the populations to allow for information exchange and exploration of different regions of the search space. To prevent falling into local solutions, a migration strategy can also be used to maintain diversity among the populations. Furthermore, some applications of ACO for dynamic optimization problems have been tabulated in **Table 2**.

4. Bibliometric analysis of ACO for dynamic optimization problems

The publication on Scopus by the query for title, abstract, and keywords with this query: *("ACO" OR "ant colony") AND ("dynamic optimization" OR "dynamic optimization" OR "dynamic environment" OR "uncertain environment" OR "time-varying environment" OR "dynamic problem"))* have been reviewed. The resulting list of publication years from 2003 to 2023 is plotted in **Figure 1**. **Figure 1** shows that the number of publications on the topic of dynamic optimization using the ACO is gradually increasing.

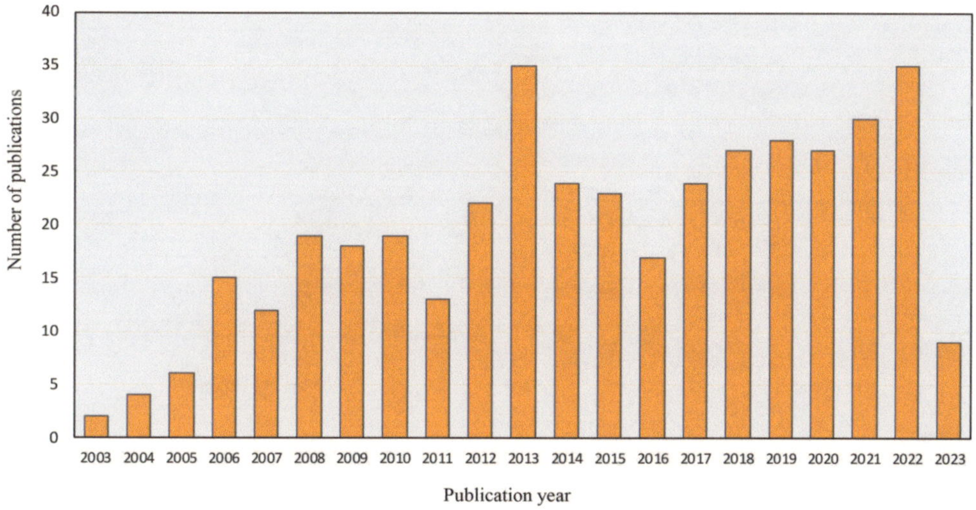

Figure 1.
Distribution of the number of publications that used ACO for dynamic optimization problems.

Furthermore, the top 10 countries with the most publications are plotted in **Figure 2**. As shown in **Figure 2**, most publications are published by researchers from China (34%).

The top 10 universities with the most publications are plotted in **Figure 3**. Looking at **Figure 3**, most publications have authors from De Montfort University, UK, having 16 records.

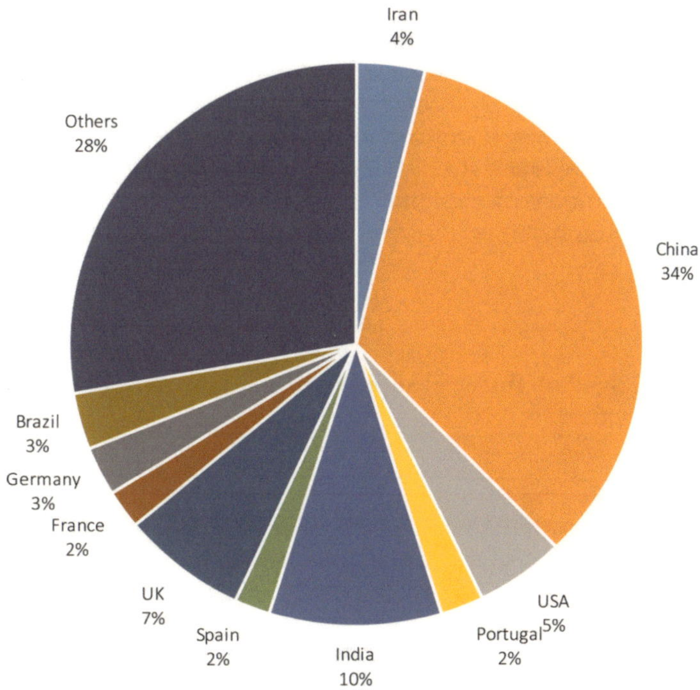

Figure 2.
Distribution of the top 10 countries with the most publications.

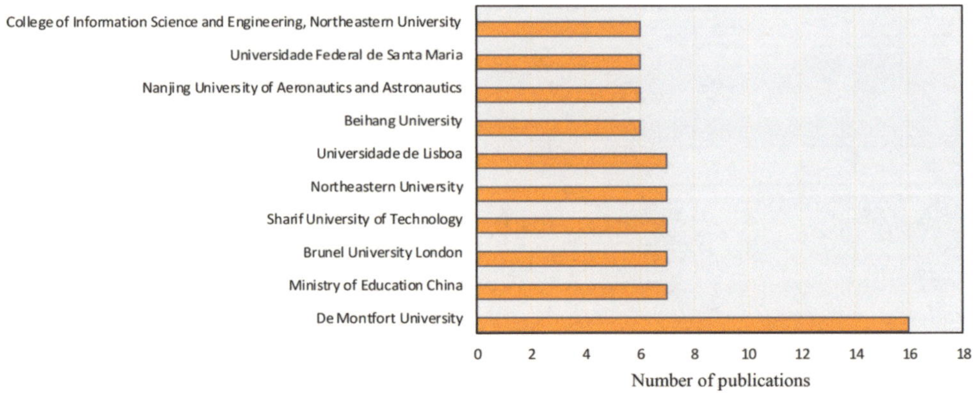

Figure 3.
Distribution of the top 10 universities with the most publications applying ACO for solving dynamic optimization problems.

5. Conclusions

As one of the most popular metaheuristics, the ant colony optimization (ACO) algorithm solves various static and dynamic optimization problems. Since most real optimization problems have dynamic and time-varying parameters with a continuous range of values, several ant colony optimization algorithms were developed to deal with dynamic optimization problems. This chapter presents a brief overview of ACO and its variants for dynamic optimization problems. Furthermore, the definition, challenges, benchmarks, measurement metrics, and a short bibliometric analysis have been provided in this chapter. In the following years, the application of ACO and its variant is expected to increase gradually for solving dynamic optimization problems.

Research can be facilitated in the future by the use of adaptive ACO algorithms, in which they can adjust the search strategy based on the current state of the environment, in order to solve dynamic optimization problems, as well as different strategies for updating pheromones, such as decaying pheromones, reinforcing pheromones, and balancing exploration with exploitation by adjusting search parameters dynamically using learning automata.

Funding

The authors received no financial support for the research, authorship, and/or publication of this article.

Conflict of interest

All authors declare that they have no conflict of interest.

Authorship change

The author names will be published exactly as they appear on the initial submission.

Ethical approval

This article does not contain any studies with human participants or animals performed by any of the authors.

Author details

Alireza Rezvanian[1]*, S. Mehdi Vahidipour[2] and Ali Sadollah[3]

1 Department of Computer Engineering, University of Science and Culture, Tehran, Iran

2 Computer Engineering Department, Faculty of Electrical and Computer Engineering, University of Kashan, Kashan, Iran

3 Department of Mechanical Engineering, University of Science and Culture, Tehran, Iran

*Address all correspondence to: rezvanian@usc.ac.ir

References

[1] Dorigo M, Di Caro G. Ant colony optimization: A new meta-heuristic. In: Proceedings of the 1999 Congress on Evolutionary Computation-CEC99 (Cat. No. 99TH8406). Washington, DC, USA: IEEE; 1999. pp. 1470-1477

[2] Pillac V, Gendreau M, Guéret C, Medaglia AL. A review of dynamic vehicle routing problems. European Journal of Operational Research. 2013; **225**(1):1-11

[3] Yang S, Cheng H, Wang F. Genetic algorithms with immigrants and memory schemes for dynamic shortest path routing problems in Mobile ad hoc networks. IEEE Transactions on Systems, Man, and Cybernetics. C Applications Review. 2010;**40**(1): 52-63

[4] Novoa-Hernández P, Corona CC, Pelta DA. Efficient multi-swarm PSO algorithms for dynamic environments. Memetic Computing. 2011;**3**:163-174

[5] Li C, Yang S, Pelta DA. Benchmark Generator for the IEEE WCCI-2012 Competition on Evolutionary Computation for Dynamic Optimization Problems. UK: Brunel University; 2011

[6] Weicker K. An analysis of dynamic severity and population size. In: Lecture Notes in Computer Science, 1917, Parallel Problem Solving from Nature-PPSN VI: 6th International Conference, Paris, France, September 18-20, 2000, Proceedings. 2000. pp. 159-168

[7] Eberhart RC and Shi Y. Tracking and optimizing dynamic systems with particle swarms. In: Evolutionary Computation, 2001. Proceedings of the 2001 Congress on Evolutionary Computation (IEEE Cat. No.01TH8546), Seoul, Korea (South). 2001. pp. 94-100.

[8] Angeline P. Tracking extrema in dynamic environments. In: Evolutionary Programming VI. Lecture Notes in Computer Science, Vol 1213. Berlin, Heidelberg: Springer; 1997. pp. 335-345

[9] Duhain JG, Engelbrecht AP. Towards a more complete classification system for dynamically changing environments. In: 2012 IEEE Congress on Evolutionary Computation. Brisbane, QLD, Australia: IEEE; 2012. pp. 1-8

[10] Branke J. Evolutionary Optimization in Dynamic Environments. New York, NY: Springer; 2002

[11] Cruz C, González JR, Pelta DA. Optimization in dynamic environments: A survey on problems, methods and measures. Soft Computing. 2010;**15**(7): 1427-1448

[12] Nguyena TT, Yangb S, Brankec J. Evolutionary dynamic optimization: A survey of the state of the art. Swarm and Evolutionary Computation. 2012;**6**: 1-24

[13] Jin Y, Branke J. Evolutionary optimization in uncertain environments-a survey. IEEE Transactions on Evolutionary Computation. 2005;**9**(3): 303-317. DOI: 10.1109/TEVC.2005.846356

[14] Yazdani D, Cheng R, Yazdani D, Branke J, Jin Y, Yao X. A survey of evolutionary continuous dynamic optimization over two decades—Part a. IEEE Transactions on Evolutionary Computation. 2021;**25**(4):609-629

[15] Yazdani D, Cheng R, Yazdani D, Branke J, Jin Y, Yao X. A survey of evolutionary continuous dynamic optimization over two decades—Part B.

IEEE Transactions on Evolutionary Computation. 2021;**25**(4):630-650

[16] Kordestani JK, Rezvanian A, Meybodi MR. An efficient oscillating inertia weight of particle swarm optimisation for tracking optima in dynamic environments. Journal of Experimental & Theoretical Artificial Intelligence. 2015;**28**:1-13. DOI: 10.1080/0952813X.2015.1020521

[17] Kordestani JK, Mirsaleh MR, Rezvanian A, Meybodi MR. Advances in Learning Automata and Intelligent Optimization. Cham, Switzerland: Springer; 2021

[18] Kordestani JK, Rezvanian A, Meybodi MR. CDEPSO: A bi-population hybrid approach for dynamic optimization problems. Applied Intelligence. 2014;**40**(4):682-694. DOI: 10.1007/s10489-013-0483-z

[19] Richter H. Detecting change in dynamic fitness landscapes. In: IEEE Congress on Evolutionary Computation. 2009. pp. 1613–1620. [Online]. Available: http://ieeexplore.ieee.org/xpls/abs_all.jsp?arnumber=4983135 [Accessed: October 30, 2012]

[20] Kazemi Kordestani J, Razapoor Mirsaleh M, Rezvanian A, Meybodi MR. An overview of multi-population methods for dynamic environments. Advances in Learning Automative Intelligence Optimization. 2021;**208**: 253-286

[21] Blackwell T. Particle swarm optimization in dynamic environments. Evolutionary Computation in Dynamic and Uncertain Environments. 2007;**51**: 29-49

[22] Branke J. Memory enhanced evolutionary algorithms for changing optimization problems. In: Proceedings of the 1999 Congress on Evolutionary Computation. Washington, DC, USA: IEEE; 1999. pp. 1875-1882

[23] Grefenstette JJ. Evolvability in dynamic fitness landscapes: A genetic algorithm approach. In: Proceedings of the 1999 Congress on Evolutionary Computation-CEC99 (Cat. No. 99TH8406), Washington, DC, USA. 1999. pp. 2031-2038

[24] Yazdani D, Omidvar MN, Cheng R, Branke J, Nguyen TT, Yao X. Benchmarking continuous dynamic optimization: Survey and generalized test suite. IEEE Transactions on Cybernetics. 2020;**52**(5):3380-3393

[25] Weicker K. Performance measures for dynamic environments. In: Parallel Problem Solving from Nature—PPSN VII. Berlin, Heidelberg: Springer; 2002. pp. 64-73. Accessed: May 22, 2014 [Online]. Available: http://link.springer.com/chapter/10.1007/3-540-45712-7_7

[26] Alba E, Sarasola B, Di Chio C. Measuring fitness degradation in dynamic optimization problems. In: Applications of Evolutionary Computatio, in Lecture Notes in Computer Sciencen. Vol. 6024. Berlin / Heidelberg: Springer; 2010. pp. 572-581. Accessed: Oct. 13, 2012 [Online]. Available: http://www.springerlink.com/index/7l6778080534516x.pdf

[27] Sarasola B, Alba E, Alba E. Quantitative performance measures for dynamic optimization problems. In: Metaheuristics for Dynamic Optimization, in Studies in Computational Intelligence. Vol. 433. Berlin/Heidelberg: Springer; 2013. pp. 17-33. Accessed: Oct. 13, 2012 [Online]. Available: http://www.springerlink.com/index/A15176531 4542484.pdf

[28] Ayvaz D, Topcuoglu HR, Gurgen F. Performance evaluation of evolutionary heuristics in dynamic environments. Applied Intelligence. 2012;**37**(1): 130-144. DOI: 10.1007/s10489-011-0317-9

[29] Kordestani JK, Rezvanian A, Meybodi MR. New measures for comparing optimization algorithms on dynamic optimization problems. Natural Computing. 2019;**18**:705-720

[30] Blackwell T, Branke J. Multiswarms, exclusion, and anti-convergence in dynamic environments. IEEE Transactions on Evolutionary Computation. 2006;**10**(4):459-472. DOI: 10.1109/TEVC.2005.857074

[31] Ranginkaman AE, Kordestani JK, Rezvanian A, Meybodi MR. A note on the paper 'A multi-population harmony search algorithm with external archive for dynamic optimization problems' by Turky and Abdullah. Information Sciences. 2014;**288**:12-14

[32] Li C, Yang S. A general framework of multipopulation methods with clustering in undetectable dynamic environments. IEEE Transactions on Evolutionary Computation. 2012;**16**(4):556-577. DOI: 10.1109/TEVC.2011.2169966

[33] Trojanowski K and Michalewicz Z. Searching for optima in non-stationary environments. In: Proceedings of the 1999 Congress on Evolutionary Computation, 1999, (CEC 99). 1999. pp. 1–5. [Online]. Available: http://ieeexplore.ieee.org/xpls/abs_all.jsp?arnumber=785498 [Accessed: October 13, 2012]

[34] Yang S, Li C. A clustering particle swarm optimizer for locating and tracking multiple optima in dynamic environments. IEEE Transactions on Evolutionary Computation. 2010;**14**(6): 959-974. DOI: 10.1109/TEVC.2010.2046667

[35] Branke J, Schmeck H. Designing evolutionary algorithms for dynamic optimization problems. Advances in Evolution Computer Theory Applications. 2003:239-262

[36] Li C and Yang S. Fast multi-swarm optimization for dynamic optimization problems. In: Fourth International Conference on Natural Computation, 2008, (ICNC'08). 2008. pp. 624–628. [Online]. Available: http://ieeexplore.ieee.org/xpls/abs_all.jsp?arnumber=4668051 [Accessed: October 13, 2012]

[37] Alba E and Sarasola B. ABC, a new performance tool for algorithms solving dynamic optimization problems. In: 2010 IEEE Congress on Evolutionary Computation (CEC). 2010. pp. 1–7. [Online]. Available: http://ieeexplore.ieee.org/xpls/abs_all.jsp?arnumber=5586406 [Accessed: October 13, 2012]

[38] Moser I, Chiong R. Dynamic function optimization: The moving peaks benchmark. Metaheuristics Dynamic Optimization. 2013;**433**:35-59

[39] Dréo J, Siarry P. An ant colony algorithm aimed at dynamic continuous optimization. Applied Mathematics and Computation. 2006;**181**(1):457-467. DOI: 10.1016/j.amc.2005.12.051

[40] Nelder JA, Mead R. A simplex method for function minimization. The Computer Journal. 1965;**7**(4): 308-313

[41] Tfaili W, Siarry P. A new charged ant colony algorithm for continuous dynamic optimization. Applied Mathematics and Computation. 2008; **197**(2):604-613. DOI: 10.1016/j.amc.2007.08.087

[42] Socha K, Dorigo M. Ant colony optimization for continuous domains. European Journal of Operational Research. 2008;**185**(3):1155-1173

[43] Montemanni R, Gambardella LM, Rizzoli AE, Donati AV. Ant colony system for a dynamic vehicle routing problem. Journal of Combinatorial Optimization. 2005;**10**:327-343

[44] Korosec P, Silc J. The differential ant-stigmergy algorithm applied to dynamic optimization problems. In: 2009 IEEE Congress on Evolutionary Computation. Trondheim, Norway: IEEE; 2009. pp. 407-414

[45] Chen J, You X-M, Liu S, Li J. Entropy-based dynamic heterogeneous ant colony optimization. IEEE Access. 2019;7:56317-56328

[46] Zhou X, Ma H, Gu J, Chen H, Deng W. Parameter adaptation-based ant colony optimization with dynamic hybrid mechanism. Engineering Applications of Artificial Intelligence. 2022;**114**:105139. DOI: 10.1016/j.engappai.2022.105139

[47] Brest J, Korošec P, Šilc J, Zamuda A, Bošković B, Maučec MS. Differential evolution and differential ant-stigmergy on dynamic optimisation problems. International Journal of Systems Science. 2013;**44**(4):663-679

[48] Guntsch M, Middendorf M. Applying population based ACO to dynamic optimization problems. In: Ant Algorithms: Third International Workshop, ANTS 2002 Brussels, Belgium, September 12–14, 2002 Proceedings 3. Berlin, Heidelberg: Springer; 2002. pp. 111-122

[49] Shi L, Zhan Z-H, Liang D, Zhang J. Memory-based ant colony system approach for multi-source data associated dynamic electric vehicle dispatch optimization. IEEE Transactions on Intelligent Transportation Systems. 2022;**23**(10):17491-17505

[50] Xu B, Cheng W, Qian F, Huang X. Self-adaptive differential evolution with multiple strategies for dynamic optimization of chemical processes. Neural Computing and Applications. 2019;**31**:2041-2061

[51] Wu L-J, Shi L, Zhan Z-H, Lai K-K, Zhang J. A buffer-based ant colony system approach for dynamic cold chain logistics scheduling. The IEEE Transactions on Emerging Topics in Computational Intelligence. 2022;**6**(6):1438-1452

[52] Liu J, Huang T, Duan X. Application of improved ant Colony algorithm in optimal planning of dynamic distribution network. In: Journal of Physics: Conference Series. IOP Publishing; 2020. p. 032068

[53] Randall M. A dynamic optimisation approach for ant colony optimisation using the multiple knapsack problem. In: The Second Australian Conference on Artificial Life. Sydney: World Scientific; 2005

[54] Sim KM, Sun WH. Ant colony optimization for routing and load-balancing: Survey and new directions. IEEE Transactions on Systems, Man, and Cybernetics - Part A: Systems and Humans. 2003;**33**(5):560-572

[55] Muteeh A, Sardaraz M, Tahir M. MrLBA: Multi-resource load balancing algorithm for cloud computing using ant colony optimization. Cluster Computing. 2021;**24**(4):3135-3145

[56] Gao S, Wang Y, Cheng J, Inazumi Y, Tang Z. Ant colony optimization with

clustering for solving the dynamic location routing problem. Applied Mathematics and Computation. 2016; **285**:149-173

[57] Eaton J, Yang S, Mavrovouniotis M. Ant colony optimization with immigrants schemes for the dynamic railway junction rescheduling problem with multiple delays. Soft Computing. 2016;**20**:2951-2966

[58] Silva CA, Runkler TA. Ant colony optimization for dynamic traveling salesman problems. ARCS 2004–Organic Pervasive Computing. 2004:259-266

[59] Xiang X, Qiu J, Xiao J, Zhang X. Demand coverage diversity based ant colony optimization for dynamic vehicle routing problems. Engineering Applications of Artificial Intelligence. 2020;**91**:103582

[60] Xiang X, Tian Y, Zhang X, Xiao J, Jin Y. A pairwise proximity learning-based ant colony algorithm for dynamic vehicle routing problems. IEEE Transactions on Intelligent Transportation Systems. 2021;**23**(6): 5275-5286